MULTIMEDIA WIRELESS NETWORKS

Technologies, Standards, and QoS

Prentice Hall PTR Communications Engineering and Emerging Technologies Series

Theodore S. Rappaport, *Series Editor*

MULTIMEDIA
WIRELESS NETWORKS

Technologies, Standards, and QoS

Aura Ganz

Zvi Ganz

Kitti Wongthavarawat

PRENTICE HALL PTR
UPPER SADDLE RIVER, NJ 07458
WWW.PHPTR.COM

Library of Congress Cataloging-in-Publication Data

A catalog record for this book can be obtained from the Library of Congress

Publisher: *Bernard Goodwin*
Editorial/production supervision: *Nicholas Radhuber*
Cover design director: *Jerry Votta*
Cover design: *Nina Scuderi*
Manufacturing manager: *Maura Zaldivar*
Editorial assistant: *Michelle Vincenti*
Marketing manager: *Dan DePasquale*

Prentice Hall books are widely used by corporations and government agencies for training, marketing, and resale.

The publisher offers excellent discounts on this book when ordered in quantity for bulk purchases or special sales. For more information, please contact:
U.S. Corporate and Government Sales
1-800-382-3419
corpsales@pearsontechgroup.com

For sales outside the U.S., please contact:
International Sales
1-317-581-3793
international@pearsontechgroup.com

ISBN 0-13-046099-0

Pearson Education LTD.
Pearson Education Australia PTY, Limited
Pearson Education Singapore, Pte. Ltd.
Pearson Education North Asia Ltd.
Pearson Education Canada, Ltd.
Pearson Educación de Mexico, S.A. de C.V.
Pearson Education—Japan
Pearson Education Malaysia, Pte. Ltd.

CONTENTS

P R E F A C E

The introduction of wireless communication is dramatically changing our lives. The ability to communicate anytime, anywhere increases our quality of lives and improves our business productivity. The recent technological developments that allow us to execute bandwidth-hungry multimedia applications over the wireless media add new dimensions to our ability to communicate. This opens an array of exciting opportunities in business, residential, healthcare, education, leisure, and many other areas. Wireless videoconferencing will connect us with business partners and family members. Remote video medical consultation will enhance care in rural areas and at the accident scene. Interactive games that include video and graphics with partners over the globe will add new dimensions not only to our leisure opportunities but also to provisioning of an effective remote learning environment.

Such opportunities are possible due to the recent technology developments in 1) user device miniaturization, which enables adequate computation power and display in small mobile handheld devices, and 2) provisioning of significantly broader wireless links for carrying multimedia traffic. Such broader links have been introduced in the wide-ranging wireless networks including very short-range personal wireless networks, short-range local area networks, and longer range metropolitan, cellular, and satellite networks. Because of nature of the wireless environment, these links are shared among many users executing multiple applications, each requiring different levels of quality of service (QoS) support. Therefore, each wireless network needs to incorporate *bandwidth mediation policies* that enable QoS support to the different multimedia applications. It is interesting to note that the wireless networks' standards do not provide the algorithms required for such bandwidth mediation policies. Such policies are the motivation for writing this book.

Wireless networks are described in numerous public domain documents produced by standard organizations in the U.S., Europe, Asia, and other continents whose members include hundreds of participants representing companies all over the world. These organizations include IEEE (Institute of Electrical and Electronics Engineers), ETSI (the European Telecommunications Standards Institute), and ITU (International Telecommunication Union). The fact that these standards are produced by such an impressive collaboration of participants and are public domain is the basis for making these standards a true, easy means for global communication. Next generations of wireless networks include Wireless Personal Area Networks (WPANs), Wireless Local Area Networks (WLANs), Wireless Metropolitan Area Networks (WMANs), and cellular and satellite networks (see Figure 1).

The provision of a quality experience for the end-users requires end-to-end QoS support in terms of bandwidth, delay, and delay jitter. In order to provide end-to-end QoS support, we need to provide QoS in the wide area network (the Internet) as well as in the wireless extensions (WPAN, WLAN, WMAN, and cellular and satellite networks).

This book is a comprehensive guide to understanding multimedia wireless networks. The book addresses the QoS problems and solutions, discussing the architecture, applications, and implementation of wireless networks, including a number of standards and proposed standards. In this book we introduce the basic QoS support mechanisms and the standards and standardization efforts in the aforementioned array of network technologies. We focus on describing the standards' signaling mechanisms that need to be incorporated in the bandwidth mediation policies developed by network designers. We hope that the reader will realize that the development of algorithms that provide QoS support for different multimedia applications is a very complex task. We also hope that the reader will understand the available signaling mechanisms for each one of the wireless standards. Using the knowledge base provided in this book the network developers can take the necessary steps into the development and implementation of QoS mechanisms within their target wireless networks.

This book is not intended to replace the standards' documents for any design purposes. Its sole intention is to help and simplify the introduction of such standards and to provide an easy tool for making an initial comparative analysis between the different approaches presented by the standards in supporting multimedia traffic. Since the standards contain hundreds and thousands of pages of detailed information, using our personal judgment, we had to omit significant details, which we felt are less useful. Thus, we strongly recommend that for a detailed analysis and design purposes, the reader should refer to the official standards' documents.

To the best of our knowledge, this is the first book that covers such a broad array of wireless networks with a focus on multimedia support. We certainly hope to continue and develop the presented topics further by including material that describes how network designers and operators use the tools provided by the standards for providing users with an enjoyable multimedia experience. Hence, we will appreciate your feedback and input via email to Aura Ganz (ganz@ecs.umass.edu).

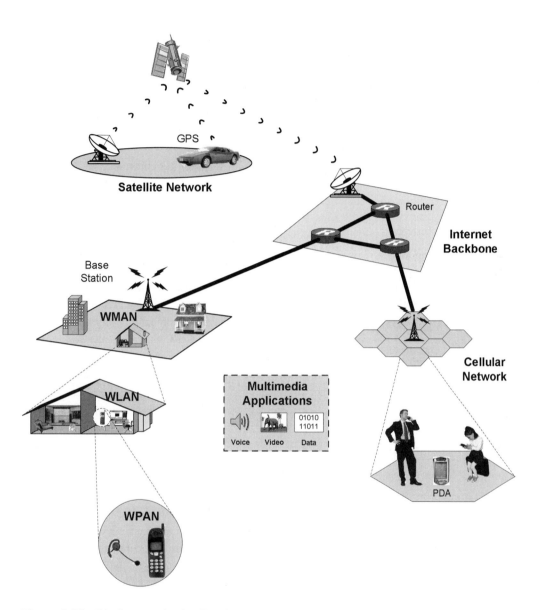

Figure 1 The Big Communication Puzzle

Target Audience

The target audience of this book includes network managers, network equipment developers, network application developers, university students, practitioners, and anyone who wants to explore QoS issues in wireless networks.

Structure of the Book

This book is organized into five primary parts.

Part 1: Multimedia Applications and Quality of Service (QoS)

In Chapter 1 we explore the nature of multimedia applications that will be delivered to the users through the wireless networks. We discuss the expectations that users have when utilizing these applications. In Chapters 2 and 3 we define QoS fundamental concepts and mechanisms and how they influence the network design. These QoS fundamentals provide the essential background required to understand the QoS support aspects of each wireless network standard in later chapters.

Part 2: Wireless Local Area Networks

In Part 2 we provide a closer look into the WLAN standards and proposed standards. Chapters 4, 5, and 6 cover IEEE 802.11, HiperLAN, and HomeRF, respectively.

Part 3: Wireless Metropolitan Area Networks

In Chapter 7 we provide a closer look into the IEEE 802.16 WirelessMAN standard.

Part 4: Wireless Personal Area Networks

In Part 4 we provide a closer look into the WPAN technologies, standards, and proposed standards. Chapters 8 and 9 describe Bluetooth and IEEE 802.15, respectively.

Part 5: 2.5G and 3G Networks

In Part 5 we provide a closer look into the 2.5G and 3G cellular networks as well as satellite network standards and proposed standards. Chapters 10, 11, and 12 cover General Packet Radio Service (GPRS), Universal Mobile Telecommunications System (UMTS), and code division multiple access (cdma2000), respectively, and Chapter 13 covers satellite networks.

Multimedia Applications and Quality of Service (QoS)

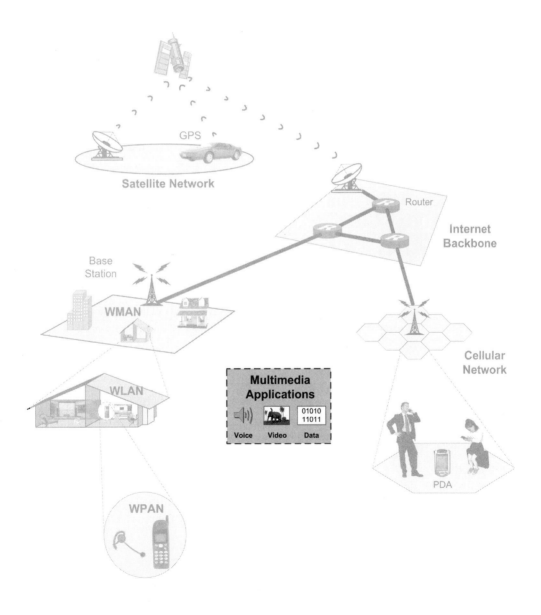

Multimedia Applications

1.1 Applications

1.1.1 Evolution

The landscape of Internet usage is dramatically changing because of the rapid evolution of access to wireless communication with its associated mobility. Consumers in both business and residential markets are becoming increasingly dependent on ubiquitous access. Such access is enabled by the proliferation of wireless communication. Consumers realize the benefits of many new business applications such as those in e-commerce, collaboration, supply chain, and tele-medicine. They also start to appreciate the benefits for their family and personal needs such as communication with friends, entertainment, gaming, and location and safety services.

As the speed and quality provisioning of wireless communication increase, consumers' dependency on applications delivered through the wireless media will increase. They will also enjoy improved graphics as the wireless media will be able to deliver content-rich applications through their higher speed wireless networks.

This new era of ubiquitous wireless communication will provide an array of applications supported by the ability to access the Internet everywhere and anytime. Wireless communication will be available at home and at the office. For example, Microsoft installed a wireless local area network (WLAN) that lets workers connect to the corporate intranet from any spot on its 265-acre Redmond, Washington, campus. Such accessibility is becoming possible also during travel via automobile, plane, ship, and train. Wireless connectivity supports Telematics, which allows drivers to access the Internet via a screen in their cars. Whether one walks in the city, suburb, or remote desert, wireless connectivity to the Internet can be established via satellites provided by companies such as Hughes Spaceway, as they can cover any point on the continent. Other types of global networks are envisioned once high-speed 3G cellular phones become available by companies such as AT&T (United States), Vodafone (United Kingdom), Orange (France), Omnitel (Italy), and DoCoMo (Japan). In the regional domain, Wireless Local Loop technologies will provide the basis for connection. WLANs provided by many companies such as 3Com, Linksys, and Cisco will cover local areas. Wireless Personal Area Networks (WPANs) provided by companies such as Ericsson will provide an option for coverage of personal areas in which communication needs to be established between neighboring devices.

Some of the applications that currently use wireless networks as their transmission media were previously available via wired media and some applications are new. Here is a partial list of these applications:

- Streaming video
- Streaming audio
- Collaboration
- One-way and interactive multimedia messaging
- Gaming, including interactive peer-to-peer (p2p) gaming
- Digital money transactions
- MP3 music download
- Video- and audio-supported shopping
- Long-distance learning, education
- Video and audio conferencing
- File sharing and transfer (pictures, video clips, and text)
- Feeding of real-time news and information about the weather, financial markets, sports and so on
- Geographic location services
- Safety services such as Enhanced 911 (E911)
- Gambling
- Entertainment

These applications will be delivered to consumers in many shapes and forms depending on the devices and wireless communication media available to the consumer.

The wireless network has a profound role in the effectiveness of the application delivery. When higher speed connections are available, the applications can deliver richer content, improved graphics, and more vivid colors. When the network can support Quality of Service (QoS), the applications can improve interactivity, reduce jitter, and provide continuous video and voice experience. When latency is high, applications that tailor location-based information for consumers may provide outdated information.

Likewise, the device architecture has a strong effect on the consumer experience. For example, a larger screen can accommodate larger pictures and more information. This dimension is amplified in the wireless environment, since devices tend to have much smaller screens. A more powerful central processing unit (CPU) can expedite computations associated with the applications, its graphics, and various communication protocols. More memory can enhance the graphic visualization experience. Moreover, such memory can be used for caching various multimedia contents for future view.

The potential list of applications that can benefit consumers in the wireless mobile environment is long and covers all aspects of life and business. We describe some applications in order to give an idea of the broad range of benefits that consumers can expect.

1.1.2 Video and Audio Streaming

Video and audio streaming provides the means of delivering news, entertainment, remote education, documentary, corporate speeches, fashion shows, and many more types of communication. Television may be the most well-known form of streaming video. It already feeds wireless multimedia streams into millions of dishes and antennas, connected to TVs and other devices. DirecTV (www.directv.com) and Dish Networks (www.dishnetworks.com) are two major providers of streaming video in the United States.

Streaming technologies are important, especially in the wireless world, since most users do not have access to enough connection capacity to download large multimedia files quickly. Using streaming technologies, consumers can start listening to the audio stream or view the video stream before the entire file has been received. To allow efficient streaming, the provider needs to send the data as a steady stream and the receiver needs to be able to cache excess data in a temporary buffer until used. If the data do not arrive fast enough, users will experience interruptions. There are several competing streaming technologies, such as RealAudio, RealVideo (www.real.com), Microsoft Media Player (www.microsoft.com), PacketVideo (www.packetvideo.com), and QuickTime (www.apple.com).

To reduce the amount of information transmitted, streaming video and audio data are compressed by means of technologies such as MPEG. The streaming video quality depends on the capacity of the transmission channel and its ability to support a steady stream—the better the channel quality (i.e., higher and steady data rate), the better the quality of the audio and video output.

1.1.3 Peer-to-Peer Computing

Peer-to-peer (p2p) computing is the sharing of files, memory, computation power, and other computer resources and services among devices. p2p Computing aggregates the shared network resources allowing economical execution of applications. For example, instead of purchasing additional memory for a PDA, the user can benefit from available memory on other devices.

In a p2p architecture devices can communicate directly with their peers without the need for a central server. They can support collaboration among users on different wireless and wired networks, moving data closer to the end user. This collaboration is obtained via caching mechanisms and distributed computing using the combined power of CPUs and memory of the devices that participate in the p2p network.

p2p Computing has an important role in both business and residential applications. Napster started the revolution in music distribution. Companies such as Groove, Endeavors, and eZmeeting are developing p2p business applications. Such applications support collaboration with rich content. Using p2p collaboration software, consumers can establish p2p work groups (see Figure 1.1) that are accessible only by predetermined or invited members. Through the click of a button, workgroup members can browse the web together, share documents, edit pictures, talk to each other through a VoIP (voice over Internet protocol) feature, share calendars, or draw diagrams on a dedicated white board.

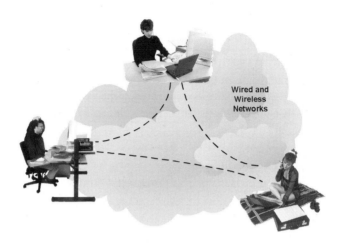

Figure 1.1   Peer-to-Peer Collaboration

1.1.4   Digital Money Transactions

Digital money transactions involving wireless devices and networks have spurred a variety of applications.

Banks will be able to support the consumer who wants to obtain cash from a traditional automated teller machine (ATM) using his or her mobile phone. To save time, the consumer is able to start the withdrawal transaction process prior to approaching the ATM. The consumer can input the account number and choose the transaction type and amount to be withdrawn. When the consumer reaches the ATM, he or she enters the security personal identification number into the mobile phone and the transaction details are transmitted to the ATM. The ATM processes the transaction and provides the amount of cash entered. Spar Nord Bank and Laan & Spar Bank in Denmark are working on such technologies.

DoCoMo and companies such as Coca-Cola introduce Cmode as part of their i-mode service. Cmode is a shopping system that uses digital transactions and that links i-mode phones to Cmode vending machines. Consumers can use their mobile phones to purchase products from Cmode-compliant machines such as vending machines which include their own computer, display screen, speakers, printers, and various sensors. To use the Cmode service, consumers need to register and deposit money into a special account maintained by DoCoMo. After the consumer makes a purchase, he or she receives a "C Ticket," which shows up on the phone's screen as a kind of bar code. When this code is passed in front of the vending machine's sensor, the machine releases the purchased product. Other communication besides the bar code transmission can be envisioned. For example, communication via a Wireless Personal Network that is established between the i-mode phone and the vending machine.

In Japan, Lawson (one of the largest chains of convenience stores with more than 7,500 shops) equipped their stores with DoCoMo iConvenience using i-mode service. With iConvenience, shoppers can make purchases through their cell phone. The shopper can enter the purchase number into the kiosk and then take the printed ticket and bar code to the cash register for payment.

1.1.5   Entertainment

One of the most prevalent applications for WLANs is entertainment. For example, home theater systems distribute movies to the monitors and speakers using wireless interconnection. The video sources (i.e., cable television, DVD player) can be in one room but the monitor can be in another room. A wireless solution makes it convenient to set up the network so people do not need to deal with wires. Digital audio files (i.e., MP3, cable television, RealAudio) have recently become popular. Instead of buying music CDs, people can purchase, download or stream digital music files from the Internet to their computers. So that the use of the entertainment is not limited to only one computer, some products provide a solution to distribute such digital music to the home stereo through a wireless audio receiver (Figure 1.2).

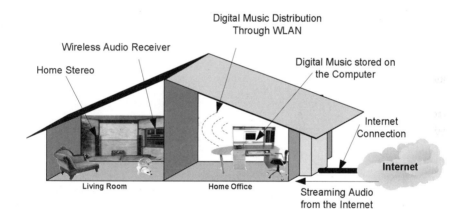

Figure 1.2   Wireless Home Entertainment

1.1.6   File and Picture Sharing

File sharing is a very common application in residential and corporate settings. Computers in the same work group or in the same domain can access and share files stored in each computer. Presently, file sharing is limited not only to a local area network, but is also extended to the Internet. There are two approaches for file sharing. One is centralized file sharing, in which there are central file servers operating as remote hard drives or online storage. A user uploads the files to a central file server. Then other users are allowed to access and download the files from the server via the Internet. For security and privacy reasons, before accessing the files, users are required to provide some form of authentication (i.e., username or password). Companies such as Xdrive (www.xdrive.com), Yahoo! Briefcase (www.briefcase.yahoo.com), and StoragePoint (www.storagepoint.net) offer such online storage services. The other type of file sharing is distributed file sharing, in which there is no central file server. The computers connecting to the Internet can share files in a p2p fashion (as described in Section 1.1.3). A computer can access and download files directly from another computer on the Internet. Software such as Kazaa Media Desktop (www.kazaa.com) and Morpheus (www.morpheus.com) enable this kind of service.

Digital pictures generated from the digital camera or picture scanner also stimulate growth of picture-sharing services. People want to share their digital pictures with friends and family. One simple solution is to append the picture as an email attachment. However, such email files become very large, which leads to long download times, especially via modem. Companies such as Clubphoto (www.clubphoto.com), ofoto (www.ofoto.com), and Yahoo! Photos (www.photos.yahoo.com) offer users the opportunity to store and share their pictures over the Internet. Some of these services also include extra features such as photo albums, slide shows, and photo printing. These services simplify the picture-sharing process, since users need to send their friends only the URLs of the remote sites.

1.1.7   Email and Multimedia Messaging

The formation of communities has been a strong desire of many consumers, especially teens. These communities are formed around enhanced messaging services that allow participants to send messages, pictures, graphics, and video and audio files to each other. Protocols such as Short Message Service (SMS), Multimedia Messaging Service (MMS), and Enhanced Messaging Service (EMS) define the architecture for various degrees of formatting, graphics, and pictures that can be exchanged among participants.

Companies such as SpotLife (www.spotlife.com) develop video communities for wireless carriers. Participants will be able to view and send personal video content to other participants. This application can also be used for the business environment. Cell phones equipped with cameras will be able to capture live video images.

1.1.8   Wireless Gaming

The wide success of hand-held gaming devices such as GameBoy and Cybiko has demonstrated the vitality of games being played on small devices in addition to large-screen televisions and personal computers. Companies such as Sega, Nintendo, and Disney have started to target mobile wireless devices as an important part of their game development strategy. Some success stories demonstrate that wireless device users have an appetite for wireless multiplayer games.

In 2001 JAMDAT Mobile (www.jamdatmobile.com) reported that 300,000 mobile device users spent more than three million air minutes in less than three months with their multiplayer game Gladiator. Gladiator is a wireless multiplayer combat game set in the Coliseum of ancient Rome. Players choose characters and weapons for their gladiators and then duel against a live opponent in real time. The more a player utilizes a particular gladiator, the more experience and skill this gladiator obtains. This company reported that players accessed their proprietary game server from more than 20 wireless carriers around the world using more than 65 different devices. Different players who participate in a multiplayer game may connect to different wireless networks with different capabilities. Their server allowed users to overcome such incompatibilities, providing a smooth gaming experience.

There are many other games that can be played over wireless networks, such as nGame's Rat Race (www.ngame.com) which allows a team of players to "race" against one another using the phone, Disney's Atlantis: The Lost Empire in which players create teams and search for the Lost Continent by sending bubbles up from the depths to get three black or three white bubbles in a row horizontally or vertically, and Disney/Pixar's Monsters Inc. in which players catch screams to achieve status as a Monsters Inc. professional kid scarer. Realizing the opportunity, Ericsson, Motorola, Nokia, Siemens, and other companies founded the Mobile Games Interoperability Forum (MGIF) (www.mgif.org). The forum's goal is to define specifications for mobile games developers so that their games can be played over various mobile devices and networks.

1.1.9   Voice and Telephony

Wireless networks that provide voice and telephony services include cellular phones, cordless phones and VoIP.

In 2002 the number of cellular phone subscribers reached one billion worldwide. The cellular phone offers not only voice service but also data services in the current 2.5G system and the upcoming 3G system. These systems allow users to surf the net or check email using their cellular phones.

VoIP provides voice or telephone service over a data network (e.g., wired LAN, IEEE 802.11 wireless LAN, and Internet). As shown in Figure 1.3, users can talk with each other via the computers connected to the data network. VoIP can be established either between computers or between computers and regular phones. An example of computer-to-computer VoIP software is MSN Messenger. Companies such as Net2Phone (www.net2phone.com) and Dialpad (www.dialpad.com) offer computer-to-phone VoIP software. An enterprise can deploy VoIP to reduce the cost of long-distance calls between branches (see Figure 1.4).

1.1.10   Location-Based Services

Location-based services are envisioned in many applications ranging from military to civilian everyday life. They include applications such as traffic reporting, restaurant recommendation, navigation, and customized ads. In such applications the information is sent to a user based on the user's location provided via the GPS (Global Positioning System).

United States military forces used it effectively in Operation Desert Storm. Supported by GPS, these forces were able to navigate through the wide, featureless desert in which the terrain looks the same for as much as the eyes can see. GPS allowed these forces to maneuver through sandstorms and during nighttime. This navigation system is also used for navigation by commercial applications. The construction of the tunnel under the English Channel was carried out simulta-

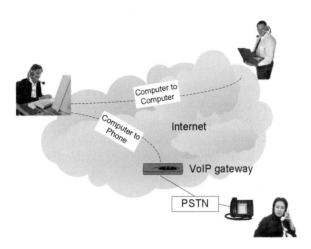

Figure 1.3   VoIP Scenario

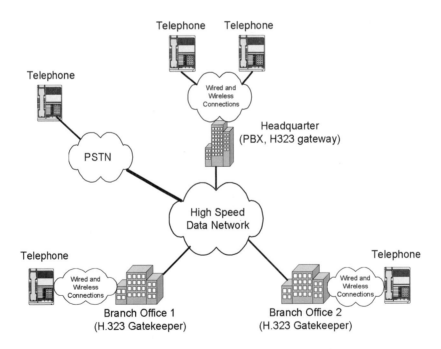

Figure 1.4   VoIP Scenario for Enterprises

neously by British and French teams digging from opposite ends of the tunnel—the English team from Dover, England, and the French team from Calais, France. These teams relied on GPS receivers outside the tunnel to validate their positions along the way and to ensure that they would meet exactly in the middle of the new tunnel. Similarly, GPS location information is being used in ships, trucks, and cars. By means of video display on the dashboard, the vehicle location data can be applied for navigation and control.

In the field of wildlife management, endangered species are being tracked by GPS receivers to monitor, study, and control population. Location information can also save human lives. One example is E911 services. They enable callers to use their cellular phones to easily place 911 emergency calls and provide the emergency services with their geographic location. In situations when the caller is traveling on an unfamiliar road, in an area with few landmarks, or at night, calling 911 on a mobile phone is probably the only choice. Using E911 services, the emergency team will have the driver's exact location, significantly shortening the time from when the rescue teams are sent until they arrive at the caller's location.

New-location based applications are invented as wireless devices equipped with GPS continue to evolve. Imagine receiving in your cell phone retail coupons based on your location. When a driver is passing a Burger King, the cell phone will beep and display an ad that suggests stopping in for one dollar off that new hamburger. The consumer can press keys to accept or refuse the offer or have an email sent for more information or future redemption. Companies such as DoCoMo (i-area service), Vindigo, and AvantGo developed such services for handheld devices. These companies report that millions of users are registered for these services.

These exciting GPS-based location services are enabled by a system of 24 satellites which are constantly orbiting the earth. They make two complete orbits every 24 hours at approximately 12,000 miles above the earth's surface. The GPS satellites continuously transmit digital radio signals that contain data on their locations and the exact time to the earth-based receivers. These satellites are equipped with atomic clocks that are very precise (less than a billionth of a second error). The receivers on earth know how long it takes for the signal to arrive. Since the signal travels at the speed of light, the receiver knows how far away the satellite is located. By using data from three satellites, the receiver computes its location in terms of longitude and latitude. By receiving a signal from one more satellite, meaning four satellites in total, the GPS receiver can also determine its altitude. The GPS system is based upon a direct line of sight between the receiver (e.g., cell phone) and the satellites. That means GPS services are not working in big cities with tall buildings or in buildings.

The FGDC (Federal Geographic Data Committee), which coordinates the development of the U.S. National Spatial Data Infrastructure, has proposed a standard for location services. This standard's objective is to increase the interoperability of location services appliances with printed maps. The committee proposes to establish a nationally consistent grid reference system as the preferred grid for National Spatial Data Infrastructure applications. This U.S. National Grid is based on universally defined coordinates and grid systems and is potentially extendable as a universal grid reference system for worldwide applications.

1.1.11 Telemedicine

Wireless communication has brought new opportunities to save lives. For example, in the remote parts of Canada, a multimedia satellite project, the Remote Communities Services Telecentre (RCST) project described in www.rcst.net, is run by a satellite operator, Telesat. RCST has been operating since 1998 linking a number of rural telecenters in Labrador and Newfoundland with others in more populated areas. These telecenters provide integrated tele-learning and telemedicine services via broadband satellite communication links. These services include high-speed Internet access, video conferencing, and digital imaging.

A physician using the RCST system in the remote Port aux Basques on the western edge of Newfoundland can, for example, transmit images of a patient's condition to a specialist in St. John's located more than 450 kilometers away. This specialist can help in diagnosing the condition, recommend treatment, or suggest a physical visit. This system saves critical time as compared to the time required for on-site diagnosis at the specialist's location, which may be especially long in difficult weather conditions.

In addition, the Integrated Emergency Medicine Network (IEMN) system has been introduced. This system provides emergency medical care for patients during ambulance transit. These special ambulances allow physicians to monitor the vital signs of patients in transit and better prepare for their arrival at the hospital (see Figure 1.5).

Similar projects have been implemented in Africa, the Philippines, and locations where the dispersed population faces difficult long-distance commutes to reach health care providers.

1.1.12 Business Applications

Business applications such as payment authorization, distance learning, multicast delivery of promotional content, and retail point-of-sale (POS) transaction processing can be handled centrally via one satellite communication provider. For example, General Motors and Ford are able to train its management and technical personnel in any office, anywhere in the world. Each student can instantly access interactive distance learning modules via DirecWay Hughes Satellites. WalMart, Shell, Texaco, and other companies use this service for their retail POS transactions.

1.1.13 Personal Area Device Connectivity

Personal area devices refer to devices carried by, placed near, or used near a person. There is a wide range of such devices:

- Consumer devices (Figure 1.6): portable CD player, digital camera, printer
- Games: game controller, joystick
- Professional devices: personal digital assistant (PDA), notebook, pager, cellular phone
- Sport training devices: health monitor, sensor, motion-tracking devices
- Hospital devices: blood pressure sensor, heart rate sensor, electrocardiogram
- Military devices: combat equipment (Figure 1.7)

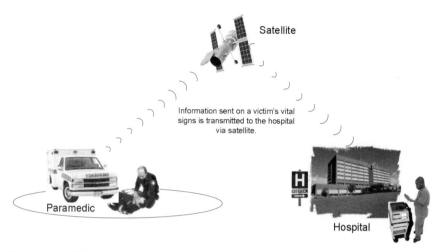

Figure 1.5   Telemedicine Scenario

Typically, personal devices have proprietary cables that interconnect (e.g., earphone, printer cable). Sometimes the cable is bulky, easily damaged, and frustrating to use. Wireless technologies become alternative solutions to replace these cables.

1.1.14 Telematics (Automobiles)

New cars are designed with more and more digital devices for every aspect including safety, operation, communication, and entertainment. There are new digital devices that control every function of the car ranging from fuel injection to airbag operation. In addition, entertainment options have increased to include satellite radio, video, and gaming. Wireless communication, combined with location services provided by GPS, has enabled safety and security services such as routing instructions, emergency door unlocking, emergency response, and stolen vehicle notification (see Figure 1.8).

Mercedes-Benz offers voice-activated phones, automatic headlights, and sensors that watch for nearby obstacles. BMW offers email access as well as its iDrive system, which allows drivers hundreds of options to access the car devices with a small control knob and on-screen menus. In addition, BMW's Mayday system offers wireless communications and GPS location services for 24-hour emergency response, roadside assistance, stolen vehicle notification, and remote door unlocking.

Similar services are offered by other companies such as General Motors. Their OnStar System integrates on-board advanced vehicle electronics with GPS technology and wireless communications technology. The OnStar System operates 24 hours a day and links the driver and car to a 24-hour response center for safety, security, and convenience needs.

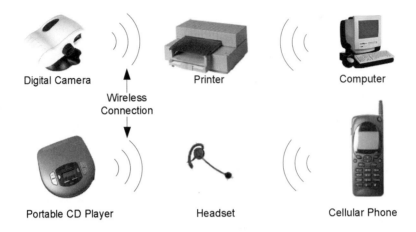

Digital Camera   Printer   Computer

Wireless
Connection

Portable CD Player   Headset   Cellular Phone

Figure 1.6   Consumer Devices

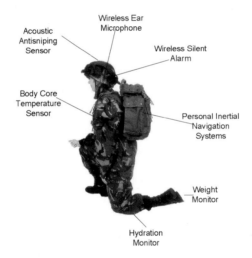

Figure 1.7   Combat Equipment

Motorola's iRadio project demonstrates wireless delivery of a wide variety of services. The iRadio converges into one package that includes picture entertainment, information, navigation, and communication. Drivers and passengers can configure their personal preferences. Its navigation system can provide real-time route planning and travel instructions, off-route notification, destination identification, traffic advisories, and local information. Radio is provided by the satellite radio channels. Users have access to news, stock quotes and other financial updates. Telephony services including email access are also supported. Internet connection can be provided through satellites or cellular network connections.

Ford, Daimler-Chrysler, General Motors, Renault, and Toyota formed the Automotive Multimedia Interface Collaboration (AMIC) forum to facilitate the development, promotion, and standardization of automotive multimedia interfaces to motor vehicle communication networks.

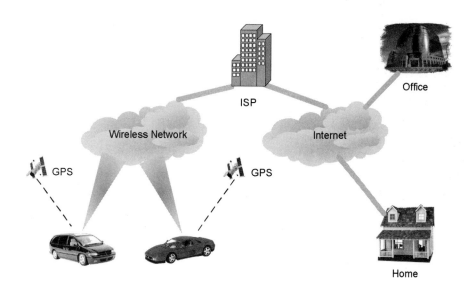

Figure 1.8 Telematics Scenario

The goal is to have an open standard that will eliminate complex and incompatible systems that can only operate on a single type of vehicle. The open standard will make it possible for electronic devices to be automatically configured and to communicate with each other. The forum considers a variety of wired and wireless technologies for the networked vehicle.

1.2 Main Protocols

1.2.1 Short Message Service (SMS)

Short Message Service (SMS) is a text message service that enables the transmission of short messages between a cell phone, PDA, PC, or any device with an Internet protocol (IP) address. Since this service was standardized in the mid 1990s, it has recorded tens of billions of messages and is being used worldwide. SMS was introduced in the Global System for Mobile Communications (GSM) and later adopted by other mobile communication systems such as Time Division Multiple Access (TDMA) and Code Division Multiple Access (CDMA).

The service supports messages that are no more than 140 to 160 alphanumeric characters in length, which is typically around 30 to 40 words in Latin and 70 characters for non-Latin alphabets like Chinese and Arabic. Initial applications of SMS were based on replacing alphanumeric pagers by supporting two-way messaging services, primarily voice mail.

SMS is used for many applications. For example, GameWorld Technologies' game entitled "Women Are Smarter?" is a mobile trivia game with a gender twist. Players respond to trivia questions that include the option to make double-or-nothing wagers on which gender answers the question correctly more often. Data regarding question and regarding which gender answers correctly more frequently are updated in real-time.

Additional services were added such as electronic mail, fax, notification service in which users are notified when predetermined events occur (e.g., receipt of email, scheduled appointments), integration of email into the SMS service, paging and its integration into the SMS service, interactive banking, and information update services (e.g., financial information, weather, news, directory assistance).

The SMS service on the SMS-enabled phone is always on—the phone is able to receive or send a short message at any time, even if a voice conversation is in progress. This is possible since the SMS messages are delivered to and from the cell phones or wireless devices over the system's control channel and not over the voice channel. This helps in preserving the voice quality while utilizing the control channel. However, the control channel capacity imposes a limit on the length of the messages since the control channel is used for control services such as phone location and call management. As the numbers of SMS messages have increased, concerns about potential service problems caused by overdemand have been mentioned.

SMS guarantees delivery of the short message. Temporary failures are identified, and the short message is stored in the network until the destination device becomes available. The message is sent to the nearest Short Message Service Center (SMSC), which either delivers it to the addressed mobile device or forwards it to the next SMSC. The messages are stored and forwarded in the SMSCs until they are received by the addressees. In the delivery process, the SMSC first sends an SMS request to the Home Location Register (HLR) to locate the addressed mobile phone. The HLR is a database used for storing and managing subscriptions and service profiles. If the addressed mobile phone is found, the HLR provides the routing information. If the HLR cannot locate the addressed phone, then the SMSC stores the message for a limited time and tries to deliver the message again when the addressee connects to the network. If the addressee is found, the message is delivered and the SMSC receives verification that the message has been received by the addressee. The sender is informed that the SMS message has been received by the addressee.

1.2.2   Enhanced Messaging Service (EMS)

Enhanced Messaging Service (EMS) evolved from the popular SMS. Service providers started to offer EMS in early 2000. EMS allows much richer content options than text-based SMS messages and can include pictures, melodies, sound marks, graphic, animations, fonts, and formatted text. For example, a person could send this text to his date with an SMS message "I love you." Using EMS, he could add a short animation of flowers along with a love melody.

EMS uses the store-and-forward mechanism provided by the SMSCs. Similar to SMS, EMS uses the control channels. EMS is enabled by SMS concatenation, linking several short messages together.

EMS supports basic and extended pictures. Basic pictures are black and white small (16x16 pixels) pictures or large (32x32 pixels) pictures. Extended pictures can be, however, black and white, grayscale, or color. Extended pictures can contain 255x255 pixels and can be transmitted in a compressed format.

EMS also supports sound. There are a few predefined sounds, including low and high chimes and chords, Claps, TaDa, and drum sounds, as well as Notify, User-Defined, and Extended Sounds. Predefined sounds are not transmitted over the air. Only a reference to the sound is included in the EMS. There are ten different sounds that can be added in the message, and when the sound reference is being displayed, the referenced sound will be played. The sender can also download melodies from various web sources. These sounds have to be formatted according to the iMelody standard. These melodies can take up to 128 bytes.

Animations in EMS provide users with a much stronger alternative for expression than plain pictures. EMS supports predefined animations that reflect happiness, sadness, flirtatiousness, gladness, skepticism, and grief. User-defined and extended animations are supported as well. Similar to sounds and pictures, EMS predefined animations are not sent as animation over the air. Only a reference to them is included in the EMS message. When the message is received by the addressee the referenced animation is displayed in a manner that is specified by the manufacturer. User-defined animations consist of four pictures. There are two different animation sizes: small (8x8 pixels) animations and large (16x16 pixels) animations. These animations are sent over the air interface. Extended animations may be black and white, grayscale, or color. The maximum size of a single animated frame is 255x255 pixels. The repetition of these animations may be controlled by the sender and can be transmitted in a compressed form.

An EMS message can be sent to a mobile phone or other wireless device, even if it does not support EMS, because all the EMS components (i.e., text formatting, pictures, animations, and sounds) are located in the message header. The EMS contents included in the header will be ignored by the receiving mobile phone if it does not support EMS, and only the text message will be displayed.

1.2.3   Multimedia Messaging Service (MMS)

Multimedia Messaging Service (MMS) technology is the ultimate messaging application, allowing users to create messages that include any combination of text, graphics, photographic images, speech, and audio or video clips. MMS supports standard image formats such as JPEG and GIF, video formats such as MPEG 4, and audio formats such as MP3 and MIDI. Multimedia messaging depends on the high transmission speeds that will be available via 3G technologies.

MMS is expected to be the future multimedia messaging technology. As compared with the SMS size limitation of 160 bytes, MMS will have no limitation even if some initial implementations may pose restrictions on the message size. Rather than sending a simple message such as "I am late," the user will be able to send a message that better explains the reasons of being late. This unlimited message size will allow users to express rich content in their messages for ultimate expression of ideas, personality, and feelings. Being able to send pictures allows users to share experiences with friends, family members, and business partners. In business applications, users can capture relevant pictures and send them to colleagues or to their home office or they can store them locally for future retrieval.

MMS built-in presentation layering can control timing and synchronization, allowing users to view, listen to, and read the messages simultaneously. MMS can support picture sharing.

MMS will be able to include still images such as pictures, screensavers, postcards, graphics, greeting cards, maps, and business cards. In addition, MMS will support animation, video, cartoons, and interactive video.

MMS will use the Wireless Application Protocol (WAP) as supporting technology and the high-speed 2.5G and 3G transmission technologies such as Enhanced Data rate for Global Evolution (EDGE), General Packet Radio Service (GPRS), and Universal Mobile Telecommunications System (UMTS). These high-speed connection technologies can provide users with the necessary bandwidth to send and receive rich-content multimedia messages. In SMS and EMS, the messages are transmitted over the control channel, which severely limits the transmission capacity. Instead, MMS will use the data channels used by all other voice and data applications. Consequently, MMS will be able to deliver much larger messages. MMS is also expected to use MExE (Mobile Execution Environment), which is a flexible and secure application environment for 2.5G and 3G mobile devices. MExE includes a variety of current technologies such as WAP and Java. Similar to SMS and EMS, MMS is a non-real-time service that routes multimedia messages to MMS servers.

MMS can include the following:

- Audio, sound, streaming audio, melodies, songs. For example, instead of sending a downloaded birthday jingle in EMS, a user can send a personalized video clip combined with the song "Happy Birthday."
- Images or pictures that are either stored or taken via a camera that is attached or part of the wireless phone or terminal. Users can take a snapshot and immediately send it to an addressee.
- Video clips. Initially the video clip may be limited to 30 seconds. This will allow users to include video clips recorded with their digital camera and to add text, labels, and appropriate audio to their messages. MMS will also support streaming video so that users can subscribe to news and entertainment services and send their own captured video as a video stream.
- Synchronized Multimedia Integration Language (SMIL, pronounced "smile") which enables simple authoring of interactive audiovisual presentations. SMIL can be used to create rich-content multimedia presentations that include streaming video and audio as well as pictures, images, text, or any other medium type. SMIL is based on an HTML-like language (standard defined by www.w3.org).

1.2.4 Wireless Application Protocol (WAP)

Wireless Application Protocol (WAP) is an open, global standard that provides a microbrowser environment optimized for wireless devices, such as phones and pagers. The goal is to easily access and interact with information over the Internet. WAP is being employed in the broad area of wireless applications described in this chapter. This standard is defined by the WAP forum whose members are wireless and Internet companies around the world to ensure interoperability and foster growth of the wireless markets. It ensures that all WAP-based applications work across all devices, from cell phones to more powerful hand-held devices. Also it allows for WAP-based applications to use less memory, processing power, display, and key handling than legacy Internet applications. Thus the WAP standard helps in delivering the multimedia content while using less bandwidth.

The most recent standard is WAP 2.0, published in 2002. It adds support for the standard Internet communication protocols: IP, TCP, and HTTP. It also allows applications to work over all existing and coming wireless 3G technologies. It provides a rich application environment, which enables delivery of information and interactive services to digital mobile phones, pagers, PDAs, and other wireless devices. It addresses the unique characteristics of wireless devices. These devices have hardware limitations (e.g., small screens, limited battery life, and limited memory) requiring special attention to user interface design (e.g., one-finger navigation).

WAP uses the Pull Model, where the user requests content from the server. WAP 2.0 adds telephony support with WTA (Wireless Telephony Application) which enables a wide range of advanced telephony applications in addition to their legacy support of data-only functionality. These tools include call handling services, such as making calls, answering them, placing them on hold, and redirecting them.

WAP 2.0 supports a Push Model, which allows server-based applications to send or "push" the content to the devices via a Push Proxy (see Figure 1.9). Push functionality is especially important for sending news, announcements, and notifications to interested users. The content can include stock prices, location-based promotions, and traffic update alerts. Without push functionality, these types of applications would require the devices to specifically pull or request application servers for such new information. In wireless environments such pulling activities, if done persistently and frequently, would burden the limited resource network with wasteful traffic.

The External Functionality Interface (EFI) service specifies the interface between the Wireless Application Environment (WAE) and plug-in modules, which extends or enhances the capabilities of browsers or other applications. The EFI framework provides for future growth and extendibility of supported WAP devices and can be used to access external devices (e.g., smart cards, GPS devices, digital cameras).

A WAP proxy (or WAP gateway) was required in the original WAP standard (see Figure 1.10). Such proxy handles the conversion between the WAP client and the origin server equipped with legacy Internet servers that are not optimized for the limited bandwidth of the wireless channel. WAP 2.0 does not require a WAP proxy. However, use of a WAP proxy can provide several important benefits such as optimizing the communication process in the wireless network and may offer mobile service enhancements such as location, secure channels (privacy), caching, and presence-based services. A proxy can also translate between WAP and other WWW protocols, allowing WAP clients to communicate with servers that do not support WAP. A WAP proxy is necessary to offer push functionality. Proxies may be located in several locations, for example, at the wireless telephone service providers and hosting companies.

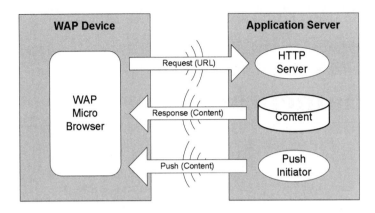

Figure 1.9 The WAP Programming Model

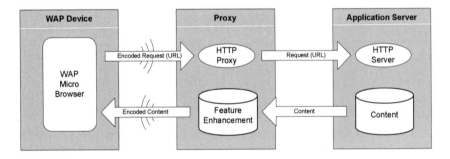

Figure 1.10 WAP Optional Proxy Model

CHAPTER 2

Quality of Service Fundamentals

2.1 Introduction

The previous chapter introduced a wide variety of multimedia applications used in wireless networks. These applications emerge to serve people's needs and sometimes create new services that attract people to deploy wireless networks in their businesses or daily lives. Multimedia applications discussed in this book mainly focus on *network applications,* i.e., applications where *hosts* and *machines* communicate through a *network.* Unlike the stand-alone multimedia applications where the multimedia contents are originated and displayed on the same machine, the network multimedia application contents that originate on a source host are transmitted through the network and displayed at the destination host. Therefore, there are a number of factors and components that affect the performance of multimedia applications such as:

- *Users:* The (human or nonhuman) ones who utilize the multimedia applications. Users' perception can influence the evaluation of the multimedia applications' performance.
- *Host machine:* The devices that operate the multimedia application (source and destination hosts). The hosts consist of a number of components such as processors, media storage systems (e.g., hard drive, CD-ROM), display devices, and operating systems.
- *Application:* The structure or the mechanisms built in the multimedia application (e.g., the codec used in video compression).
- *Network:* The network components that transport the multimedia contents between the two host machines (source and destination). Examples of hardware components include: switches, routers, network interface cards, gateways, and firewalls. The network also includes network protocols that reside at each network hardware element.

Figure 2.1 shows a simplified diagram of a communication system that includes all the components mentioned above. Hosts connect to the network through the network interface devices using various network technologies (e.g., IEEE 802.11 WLAN, Satellite, Bluetooth). The network cloud consists of multiple segments of interconnected subnetworks that establish the communication path between the hosts. All of these components require multimedia support.

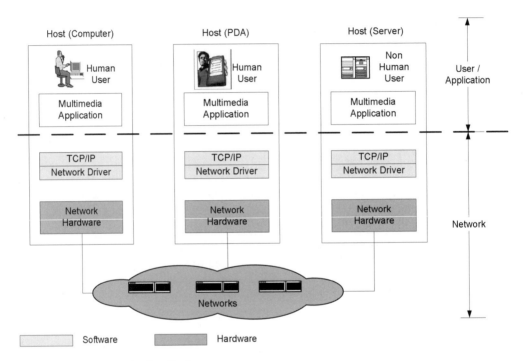

Figure 2.1   Communication System

Multimedia support issues can be presented by using the Quality of Service (QoS) term, which is an overloaded term with various meanings and perspectives. There is little consensus on the precise definition of QoS. Different people and communities perceive and interpret QoS in different ways. For example, in the networking community, QoS refers to the service quality or service level that the network offers to applications or users in terms of network QoS parameters, including latency or delay of packets traveling across the network, reliability of packet transmission, and throughput. However, in application communities, QoS generally refers to the application quality as perceived by the user—that is, the presentation quality of the video, the responsiveness of interactive voice, and the sound quality (CD-like or FM-radio-like sound) of streaming audio. Instead of providing a precise QoS definition, we present a simplified QoS model that includes two QoS perspectives: Application/User and Networks (see Figure 2.2).

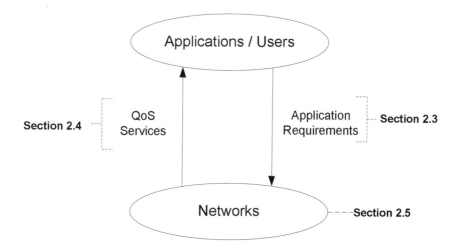

Figure 2.2   QoS Model and Corresponding Section in This Chapter

We assume that applications and users are in the same group because of their close relationship and the common way they perceive quality. In this book we often use the terms "applications" and "users" interchangeably. The applications/users expect a specific QoS in terms of response time, for example consistent perceived quality and uninterrupted service. These QoS requirements are passed to the network implicitly or explicitly and the underlying networks are responsible in part for meeting these requirements.

Users are not concerned about how the network manages its resources or what mechanisms are involved in QoS provision. However, the users are concerned about the services that networks provide which directly impact the perceived quality of the application.

From the network perspective, the networks' goal is to provide the QoS services that adequately meet the users' needs while maximizing the network resources (i.e., bandwidth) utilization. To achieve this goal, the networks analyze the application requirements, manage the network resources, and deploy various network QoS mechanisms.

Although the QoS model here looks rather simple, there are a number of subtle issues that need to be discussed. For example:

- What information is contained in the application requirements?
- Desired quality of the application which relates to users' satisfaction is subjective. How does the desired quality map to QoS parameters which are managed by networks? What criteria need to be considered in the mapping process?
- What are the QoS parameters?
- Which QoS mechanisms do the networks use to achieve diverse QoS support?
- What kind of QoS service level can networks offer?
- What is the relationship between QoS services and application requirements?

In the remaining part of this chapter we attempt to answer these issues. This chapter is organized as follows. In Section 2.2 we introduce the application and network QoS parameters. Section 2.3 discusses multimedia application requirements in terms of QoS parameters. QoS services are introduced in Section 2.4, and the realization of QoS services is described in Section 2.5.

2.2 QoS Parameters

The following QoS parameters are relevant to multimedia applications:

1. Throughput or bandwidth
2. Delay or latency
3. Delay variation (delay jitter)
4. Loss or error rate

2.2.1   Throughput

From the application perspective, throughput refers to the data rate (bits per second) generated by the application. Throughput, measured in the number of bits per second, sometimes is called *bit rate* or *bandwidth*. Bandwidth is considered to be the network resource that needs to be properly managed and allocated to applications.

The throughput required by an application depends on the application characteristics. For example, in a streaming video application, different video properties generate different throughput. A user can select the video quality by varying the following video properties:

- *Frame size:* A function of the number of pixels in each row and column and of the number of bits per pixel.
- *Frame rate:* The refreshing video frame rate (number of frames per second). Decreasing video frame rate reduces the bandwidth consumption but compromises the smoothness of the video movement.
- *Color depth:* The number of possible colors represented by a pixel. The 256-color video requires 8 bits of data per pixel; whereas, the 16-million-color video requires 24 bits of data per pixel.
- *Compression:* Reduction of the bandwidth consumption at the expense of image quality. Examples of video compression standards include MPEG1, MPEG2, and MPEG4.

In the remaining part of this subsection we will describe two characteristics of an application data traffic generation process: the data traffic generation rate (constant and variable bit rate) as well as data traffic generation burstiness. We will show how the traffic generation process determines the applications' required throughput.

2.2.1.1   Constant and Variable Bit Rate

Constant bit rate (CBR) applications generate data traffic with constant data rate. Examples of CBR applications include the following:

- Digital telephone Private Branch Exchange (PBX), which generates 64 kbps constant bit rate
- Uncompressed digital video

Most of the constant bit rate applications are delay sensitive and require constant bandwidth allocation. Allocating bandwidth below the required bandwidth causes application failure. On the other hand, allocating bandwidth above the requirement does not improve the user satisfaction, as shown in Figure 2.3.

Variable bit rate (VBR) applications generate data traffic with variable data rate. The degree of bit rate variability depends on the application. Examples of VBR applications include the following:

- Compressed video and audio
- Remote login

VBR applications require minimum bandwidth allocation in order to operate successfully. The more allocated bandwidth, the better the user-perceived quality. Bandwidth allocation beyond the maximum required bandwidth does not improve user satisfaction as shown in Figure 2.3.

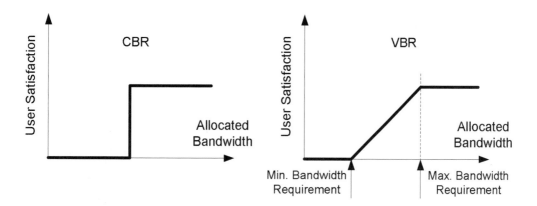

Figure 2.3   User Satisfaction as a Function of Allocated Bandwidth for CBR and VBR traffic

2.2.1.2   Burstiness

The data traffic *burstiness* measures the degree of bit rate variability of a VBR application. The burstiness is defined as the ratio between the Mean Bit Rate (MBR) and Peak Bit Rate (PBR) where:

- PBR is the maximum number of bits in a short period of time
- MBR is the average number of bits in a long period of time

2.2.2   Delay

Delay has a direct impact on users' satisfaction. Real-time applications require the delivery of information from the source to the destination within a certain period of time. Long delays may cause incidents such as data missing the playback point, which in turn reduces the video fidelity. Moreover, it can cause user frustration during interactive tasks. When the data traffic is carried across a series of components in the communication system that interconnects the source and the destination, each component introduces delay. We can categorize the main sources of delay as follows:

1. *Source-processing delay* (digitization and packetizing delay): This delay, which is introduced by the source that generates the packets, depends on the source host hardware configuration (CPU power, RAM, motherboard, etc.) and its current load (e.g., the number of applications running simultaneously and their required hardware resources).
2. *Transmission delay:* The transmission time of a packet is a function of the packet size and transmission speed.
3. *Network delay:*
   a. *Propagation delay:* The propagation delay from the source to the destination is a function of the physical distance between the source and the destination.
   b. *Protocol delay:* The delay is caused by the communication protocols executed at the different network components such as routers, gateways, and network interface cards. The delay depends on the protocols, the load of the network, and the configuration of the hardware that executes the protocol.
   c. *Output queuing delay:* The delay is caused by the time a packet spends in the outgoing link queue at a network component. For example, such delay can be incurred at an intermediate router output queue. The delay depends on the network congestion, the configuration of the hardware, and the link speed.
4. *Destination processing delay:* This delay is introduced by the processing required at the destination. For example, such delay can be incurred in the packet reconstruction process. Similar to the processing delay at the source, this delay depends on the destination host hardware configuration and load.

Figure 2.4 depicts the end-to-end delay diagram.

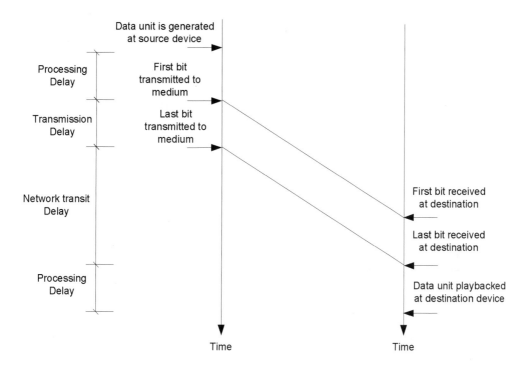

Figure 2.4 End-to-end Delay Diagram

2.2.3 Delay Variation

Delay variation is a QoS metric that refers to the variation in the delay introduced by the components along the communication path. Since each packet in the network travels through different paths, and the network conditions for each packet can be different, the end-to-end delay varies. For data generated at constant rate, the delay jitter distorts the time synchronization of the original traffic. As shown in Figure 2.5 the packets travel through the network and experience different end-to-end delays, reaching the destination with timing distortions (incomplete or delayed signal) relative to the original traffic. There are several techniques to cope with delay jitter at the receiver end. For example, in technique A (Figure 2.5A), the receiver playbacks the signal as soon as the packets arrive. The playback point is changed from the original timing reference. This introduces distortion in the playback signal. In technique B (Figure 2.5B), the receiver playbacks the signal based on the original timing reference. The late packets that miss the playback point will be ignored. This also introduces distortion. In technique C (Figure 2.5C) a de-jittered buffer is used. All packets will be stored in the buffer and held for some time (offset delay) before they are retrieved by the receiver with the original timing reference. The fidelity of the signal will be maintained as long as there are packets available in the buffer. Large delay jitter requires large buffer space to hold the packets and smooth out the jitter. A large buffer also

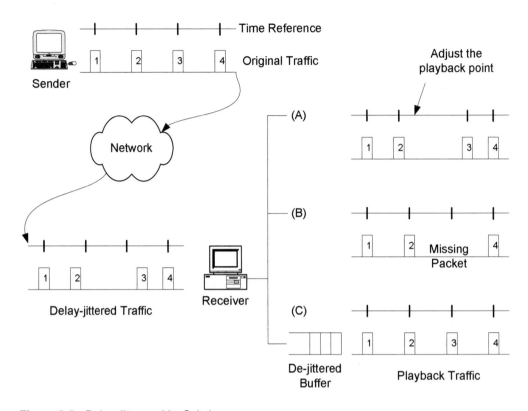

Figure 2.5   Delay Jitter and Its Solutions

introduces large delays, which will be eventually constrained by the application delay require-ment. In summary, there is a tradeoff between the following three factors: de-jittered buffer space, delay requirement, and fidelity of the playback signal.

2.2.4   Loss or Error Rate

Packet loss directly affects the perceived quality of the application. It compromises the integrity of the data or disrupts the service. At the network level, packet loss can be caused by network congestion, which results in dropped packets. Another cause of loss is caused by bit errors that occur due to a noisy communication channel. Such loss will most likely occur in a wireless channel. There are several techniques to recover from packet loss or error such as packet retrans-mission, error correction at the physical layer, or codec at the application layer that can compen-sate or conceal the loss.

2.3   Multimedia Application Requirements

Different multimedia applications have different QoS requirements expressed in terms of the following QoS parameters as described in the previous section: throughput, delay, delay variation, and loss. In many cases, users can determine the application's QoS requirements by investigating the factors that influence the application quality (i.e., the task characteristics, user characteristics). For example, from experimentation they conclude that for acceptable quality, the one-way delay requirements of interactive voice should be less than 250 ms. This delay value includes the delay introduced from all components in the communication path such as source delay, transmission delay, network delay, and destination delay.

In this book, we focus on the applications' requirements for multimedia support in wireless networks. Due to advances in the source and destination available processing power and the very high-speed optical infrastructure, we expect delay and bandwidth bottlenecks to occur in the wireless network due to their limited available bandwidth.

Before presenting detailed QoS requirements for each application, we would like to provide an overview of the factors that influence the application requirements. Some of the factors are as follows:

- *Application interactivity level:* Interactive and noninteractive applications
- *User/Application characteristics*: Delay tolerance and intolerance, adaptive and nonadaptive characteristics
- *Application criticality:* Mission-critical and non-mission-critical applications

We will describe below each one of these factors.

2.3.1   Interactive and Noninteractive Applications

An interactive application involves some form of interaction (action-reaction, request-response or exchange of information) between two parties (people-to-people, people-to-machine or machine-to-machine). Examples of interactive applications include the following:

- *People-to-people application:* IP telephony, interactive voice/video, videoconferencing
- *People-to-machine application:* Video-on-demand (VOD), streaming audio/video, virtual reality
- *Machine-to-machine application:* Automatic machine control

The elapsed time between interactions is essential to the success of an interactive application. The degree of interactivity determines the level or stringency of the delay requirement. For example, interactive voice applications, which involve human interaction (conversation) in real time, have strict delay requirements (in the order of milliseconds). Streaming (playback) video applications involve less interaction, (i.e., interaction mostly occurs during start, stop, forward, or reverse action on the video) and do not require real-time response. Therefore streaming applications have more relaxed delay requirements (in the order of seconds). Often the applications' delay tolerance is determined by the users' delay tolerance (i.e., higher delay tolerance leads to more relaxed delay requirements).

Delay jitter is also related to QoS support for interactive tasks. As discussed above, the delay jitter can be corrected by de-jittering buffer techniques. However, the buffer introduces delay in the original signal, which eventually affects the interactivity of the task. In general, an application with strict delay requirements also has strict delay jitter requirements.

2.3.2   Tolerance and Intolerance

Tolerance and intolerance describe the users' sensitivity to changes in QoS parameter values. We next describe users' tolerance to latency and distortion.

- *Latency tolerance and intolerance:* This characteristic determines the stringency of the delay requirement. As we have discussed above, streaming multimedia applications are more latency tolerant than interactive multimedia applications. The degree of latency tolerance depends on users satisfaction, users expectation, or the urgency of the application (i.e., remote machine control in the manufacturing production line is latency intolerant).
- *Distortion tolerance and intolerance:* The tolerance to the fidelity of the application quality depends on factors such as users' satisfaction, users' expectation, and the application media types. For example, users are more tolerant to video distortion than to audio distortion. In this case, during congestion, the network has to maintain the quality of the audio output over the quality of the video output.

2.3.3   Adaptive and Nonadaptive Characteristics

Adaptive and nonadaptive aspects mostly describe the mechanisms invoked by the applications to adapt to QoS degradation. The common adaptive techniques are rate adaptation and delay adaptation:

- *Rate adaptive application* can adjust the data rate injected into the network. During network congestion, the applications reduce the data rate by dropping some packets, increasing the codec data compression, or changing the multimedia properties (i.e., reducing the resolution or color depth of the video). These schemes will most likely cause degradation of the perceived quality but will keep it within acceptable levels.
- *Delay-tolerant adaptive applications* tolerate a certain level of delay jitter by deploying the de-jittered buffer or adaptive playback technique (Figure 2.5).

Adaptation is trigged by some form of implicit or explicit feedback from the network or end user. Again, like tolerance, adaptive applications provide some room for networks to readjust the QoS services.

2.3.4 Application Criticality

Mission-critical and non-mission-critical aspects reflect the importance of application usage, which determines the strictness of the QoS requirements. Failing the mission may result in disastrous consequences. For example:

- *Remote surgery:* The surgeon performs an operation through remote surgical equipment. Life and death of the patient may depend on the promptness and accuracy of the surgical equipment control.
- *Telemedicine:* The accuracy of medical images (i.e., magnetic resonance image, x-ray image, ultrasound image) is extremely important. Distorted images may lead to wrong diagnosis.

2.3.5 Representation of Application Requirements

There are two common ways to express the applications' requirements: quantitative expression and qualitative expression.

2.3.5.1 Quantitative Expression

Application requirements are expressed in terms of QoS parameters with quantifiable values that can be determined from the application's technical specifications (i.e., video codec—MPEG 1, MPEG2, MPEG4, HDTV) or from experimentation. Some applications may obtain these values using runtime measurements.

- *Throughput:* It is mostly expressed as the average data rate:

  - Uncompressed HDTV: 1.5 Gbps
  - MPEG4: 5 kbps – 4 Mbps
  - ITU-T G.711: 64 kbps
  - ETSI GSM 06.10: 16 kbps

  For applications that generate VBR traffic, the average data rate does not properly capture the traffic characteristics. Burstiness parameters (average rate, peak rate, maximum burst size) may also be included.

- *Delay and delay jitter:* The value is provided in the form of a bound (i.e., the delay should be less than a certain value). In mission-critical or delay intolerant applications, the network has to follow the delay requirement strictly. In delay-tolerant applications, the bound value is the average value. The network can provide the service in a more relaxed manner (i.e., some of the packets can miss their deadline).
- *Loss:* It is mostly expressed as a statistical value (i.e., the percentage of lost packets should be less than a certain value).

2.3.5.2   Qualitative Expression

In contrast to the quantitative expression, QoS requirements are expressed qualitatively—for example, "…. get the bandwidth as much as you can give… " or " … expect low latency…" or " …. provide lower latency than certain applications…." There are a number of reasons why some applications (e.g., web browsing, remote login) express their QoS requirements qualitatively: some applications only need better service than others, or some applications are incapable of quantifying their QoS requirements, or it is too costly to quantify the QoS requirements. Since these applications do not specify quantitative QoS requirements, it is intractable for networks to provide quantitative services.

2.3.6   Examples of Application Requirements

2.3.6.1   Interactive Voice

Requirements for voice applications such as voice over IP (VoIP) can be exemplified as follows:

- *Bandwidth:* VoIP requires relatively low bandwidth (i.e., 64 kbps in ITU G.711) and the traffic pattern is relatively constant. Voice traffic is packetized in small packets of around 44 to 200 bytes resulting in short transmission times. Since voice conversations contain up to 60% silence, voice encoding algorithms yield lower bandwidth requirements with acceptable quality. Table 2.1 shows the speech coding standard. Mean Opinion Score (MOS) in Table 2.1 reflects the sound quality in a quantitative way. MOS is an empirical value determined by averaging the opinion score (i.e., the range from 1=worst to 5=best) of a sample group of listeners who listen and evaluate the voice quality.

Table 2.1   Speech Coding Standards

Codec	Bandwidth (kbps)	Sound Quality (MOS)	Codec Complexity
ITU-T G.711	64	> 4	—
ITU-T G.722	48-64	3.8	low
ITU-T G.723.1	6.4/5.3	3.9	high
ITU-T G.726	32	3.8	low
ITU-T G.728	16	3.6	low

Table 2.1 Speech Coding Standards (Continued)

Codec	Bandwidth (kbps)	Sound Quality (MOS)	Codec Complexity
ITU-T G.729	8	3.9	medium
GSM 06.10	13	3.5	low
GSM 06.20	5.6	3.5	high
GSM 06.60	12.2	> 4	high
GSM 06.70	4.8 – 12.2	> 4	high

- *Delay:* Since VoIP involves human interaction (conversation), human perception is sensitive to both the sound quality and the conversation gap and response. Table 2.2 displays delay guidelines for VoIP.

Table 2.2 Delay Guidelines for VoIP

One-Way Delay	Effect on Perceived Quality
< 100 – 150 ms	Excellent quality (undetectable delays)
150 – 250 ms	Acceptable quality (slight delays)
250 – 300 ms	Unacceptable quality

- *Delay jitter:* A large delay jitter causes delayed packets which impact the quality of the voice (conversation gap). Table 2.3 shows the delay jitter guidelines for VoIP.

Table 2.3 Delay Jitter Guidelines for VoIP

Delay Jitter	Effect on Perceived Quality
< 40 ms	Excellent quality—Undetectable jitter
40—75 ms	Acceptable quality
Over 75 ms	Unacceptable quality

- Not only the packet loss has impact on the voice quality, but also the pattern of the packet loss has impact. For example, the loss of one packet may not be noticeable due to the sophisticated codecs that can conceal the loss. However, two or more consecutive lost packets can cause voice quality deterioration.

2.3.6.2 Video Applications

Video applications can be classified into two groups: interactive video (i.e., video conferencing, long-distance learning, remote surgery) and streaming video (i.e., RealVideo, Microsoft ASF, QuickTime, Video on Demand, HDTV). As shown in Table 2.4, video applications' bandwidth requirements are relatively high depending on the video codec.

Table 2.4   Video Codec Bandwidth Requirements

Video Codec	Bandwidth
Uncompressed HDTV	1.5 Gbps
HDTV	360 Mbps
Standard definition TV (SDTV)	270 Mbps
Compressed MPEG2 4:4:4	25 – 60 Mbps
Broadcast quality HDTV (MPEG 2)	19.4 Mbps
MPEG 2 SDTV	6 Mbps
MPEG 1	1.5 Mbps
MPEG 4	5 kbps – 4 Mbps
H.323 (H.263)	28 kbps – 1 Mbps

- *Interactive video applications:* Interactive video which involves human interaction requires low end-to-end delay. Low delay jitter is also required especially in mission-critical applications such as remote surgery. Interactive video users are more tolerant to video distortion than streaming video users.

- *Streaming video applications:* Streaming video is considered to be one-way communication (from video server to users). The only interaction between the server and the user occurs during start, stop, forward, or reverse commands. Therefore, the delay is not a main concern. However, this application is delay jitter intolerant. Large delay jitter causes packets to miss the playback point, causing video distortion.

2.4 QoS Services

Networks receive from the applications (implicitly or explicitly) their QoS requirements through quantitative or qualitative expression. Networks need to respond to these requests by supplying QoS services using a number of QoS mechanisms. In this section, we describe various types of QoS services and in the next chapter we will introduce the mechanisms required to provide such services. In the rest of the book we will analyze each wireless technology and discuss which kind of services it can provide and which kind of applications it can support.

We categorize the QoS services as follows:

1. What kind of service is provided to applications: *quantitative services, qualitative services, best effort services*
2. To which entities (individual or group [class] of applications) the network provides service: *per-flow QoS services, per-class QoS services.*

Networks may use a combination of QoS services (i.e., per-flow and quantitative, per-class and quantitative). Some networks may include multiple types of QoS services in order to support a wide range of applications. Example of QoS services include the following:

- *Guaranteed Integrated Services (IntServ),* which provide per-flow and quantitative QoS service
- *Controlled Load Integrated Services,* which provide per-flow and qualitative QoS service
- *Differentiated Services (DiffServ),* which provide per-class and qualitative QoS service

We next describe in more detail each one of these QoS services.

2.4.1 Quantitative (Guaranteed) Services

Quantitative (guaranteed) services, or services that deliver hard QoS, guarantee the provision of the application quantitative requirements. This service delivers the highest quality of service. The guaranteed services guarantee the network performance (i.e., bandwidth, delay, delay jitter) in deterministic or statistical terms. For example, networks guarantee the minimum bandwidth provided to an application or guarantee the delay bound of packet delivery within a certain value. The services are suitable for applications that require quantitative performance guarantee such as mission-critical and interactive applications. As described in the next chapter, a number of QoS mechanisms are required to enable these services.

2.4.2   Qualitative (Differentiated) Services

Qualitative (differentiated) services, or services that deliver soft QoS, provide relative services. In other words, qualitative services may provide lower delay to one class of applications than to another class of applications. One prominent differentiated services example is the priority service. An application that belongs to a higher priority class will receive service before applications that belong to a lower priority class.

2.4.3   Best Effort Services

Best effort services provide network services without any performance guarantees. All traffic is treated equally. Best effort services are suitable for data traffic (i.e., FTP, email, web pages) that does not require minimum bandwidth or timed delivery.

2.4.4   Per-Flow QoS Services

Per-flow QoS services provide service assurance to individual flows (applications) quantitatively or qualitatively. For example, interactive video has stricter delay requirements than streaming video. Therefore, the network has to provide different services to each application in order to meet its individual needs. Per-flow classification (QoS mechanism that differentiates the flows) is essential to the implementation of per-flow QoS services.

2.4.5   Per-Class QoS Services

Applications are categorized into different groups (classes) based on different criteria (i.e., QoS requirement, organization, application types, protocol families). Per-class QoS services provide service assurance to individual classes quantitatively or qualitatively. Applications in the same class will experience the same QoS. The per-class classification (QoS mechanism that identifies and differentiates the collective entities) is essential to the implementation of per-class QoS services.

2.5   Realization of QoS Services

There are a large number of approaches, mechanisms, and technologies deployed in the network in order to enable the QoS services introduced in the previous section. The bandwidth, which is the main network resource, needs to be distributed to all applications in a way that simultaneously satisfies all QoS requirements. To enable QoS services, there are two main approaches (philosophies) based on how they deal with bandwidth planning in order to enable QoS services:

- *Bandwidth over-provisioning:* When the current network bandwidth cannot provide QoS support, the network infrastructure is upgraded. Using a higher bandwidth infrastructure may result in less congestion and therefore lower delivery delays.
- *Bandwidth management:* This approach proposes to manage the bandwidth using QoS mechanisms. Examples of QoS mechanisms which will be described in the next chapter are classification, admission control, resource reservation, channel access, packet scheduling, and policing.

There is an endless argument between the implementation of these two approaches. In wired networks, where bandwidth is abundant, over-provisioning is the winning approach. However, in wireless networks where the bandwidth is limited (spectrum is unavailable or is too expensive to purchase) bandwidth management techniques have to be deployed.

CHAPTER 3

QoS Mechanisms

3.1 Introduction

In the previous chapter, we introduced the fundamental QoS concepts. In this chapter we introduce a number of key QoS mechanisms that enable QoS services. At the end of this chapter, we provide a general framework for analyzing the QoS support of each wireless technology presented in the rest of this book.

QoS mechanisms can be categorized into two groups based on how the application traffic is treated: 1) traffic handling mechanisms, and 2) bandwidth management mechanisms (see Figure 3.1).

Traffic handling mechanisms (sometimes called In-traffic mechanisms) are mechanisms that classify, handle, police, and monitor the traffic across the network. The main mechanisms are: 1) classification, 2) channel access, 3) packet scheduling, and 4) traffic policing.

Bandwidth management mechanisms (sometimes called Out-of-traffic mechanisms) are mechanisms that mange the network resources (e.g., bandwidth) by coordinating and configuring network devices' (i.e., hosts, base stations, access points) traffic handling mechanisms. The main mechanisms are: 1) resource reservation signaling and 2) admission control.

3.2 Classification

The lowest service level that a network can provide is best effort service, which does not provide QoS support. In best effort service, all traffic is handled equally regardless of the application or host that generated the traffic. However, some applications need QoS support, requiring better than best effort service such as differentiated or guaranteed service. For a network to provide selective services to certain applications, first of all, the network requires a classification mechanism that can differentiate between the different applications. The classification mechanism identifies and separates different traffic into flows or groups of flows (aggregated flows or classes). Therefore, each flow or each aggregated flow can be handled selectively.

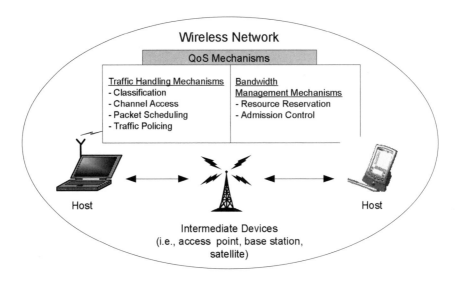

Figure 3.1   QoS Mechanisms in a Wireless Network

The classification mechanism can be implemented in different network devices (i.e., end hosts, intermediate devices such as switches, routers, access points). Figure 3.2 shows a simplified diagram of a classification module that resides on an end host and on an intermediated device.

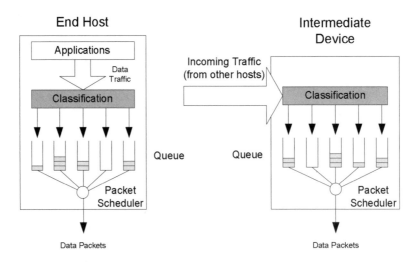

Figure 3.2   Classification

Application traffic (at the end host) or incoming traffic from other hosts (at the intermediate device) is identified by the classification mechanism and is forwarded to the appropriate queue awaiting service from other mechanisms such as the packet scheduler. The granularity level of the classification mechanism can be per-user, per-flow, or per-class depending on the type of QoS services provided. For example, per-flow QoS service requires per-flow classification while per-class QoS service requires per-class classification.

To identify and classify the traffic, the traffic classification mechanism requires some form of tagging or marking of packets. There are a number of traffic classification approaches. Some of approaches are suitable for end hosts and some for intermediate hosts. Figure 3.3 shows an example of some traffic classification approaches which are implemented in the different Open System Interconnection (OSI) layers.

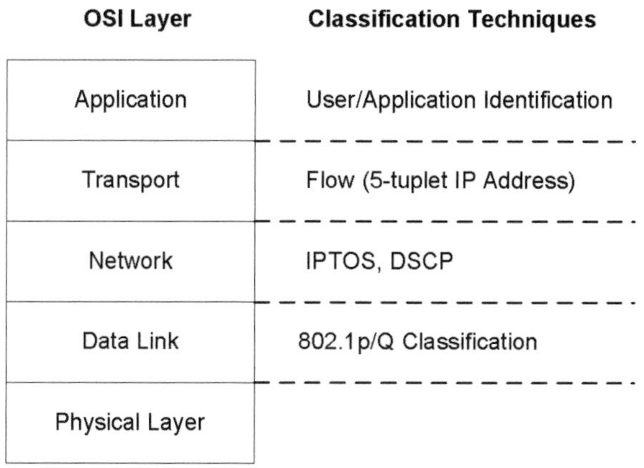

OSI Layer	Classification Techniques
Application	User/Application Identification
Transport	Flow (5-tuplet IP Address)
Network	IPTOS, DSCP
Data Link	802.1p/Q Classification
Physical Layer	

Figure 3.3   Examples of Existing Classification on Each OSI Layer

3.2.1   Data Link Layer Classification

Data link layer, or Layer 2, classifies the traffic based on the tag or field available in Layer 2 header.

An example of Layer 2 classification is IEEE (Institute of Electrical and Electronics Engineers) 802 user priority. The IEEE 802 header includes a 3-bit priority field that enables eight priority classes. It aims to support service differentiation on a Layer 2 network such as a LAN. The end host or intermediate host associates application traffic with a class (based on the Policy, or the service that the application expects to receive) and tags the packets' priority field in the IEEE 802 header. A classification mechanism identifies packets by examining the priority field of the IEEE 802 header and forwards the packets to the appropriated queues. IEEE recommends mapping the priority value and the corresponding service as shown in Table 3.1.

Table 3.1   Example of Mapping between Priority and Services

Priority	Service
0	Default, assumed to be best effort service
1	Less than best effort service
2	Reserved
3	Reserved
4	Delay sensitive, no bound
5	Delay sensitive, 100ms bound
6	Delay sensitive, 10ms bound
7	Network control

3.2.2   Network Layer Classification

Network layer, or Layer 3 classification, classifies packets using Layer 3 header. Layer 3 classification enables service differentiation in Layer 3 network.

An example of Layer 3 classification is IPTOS (Internet protocol type of service), DSCP (Internet protocol differential service code point). IPv4 and IPv6 standard defined a prioritization field in the IP header which can be used for Layer 3 classification. RFC 1349 defined a TOS field in IPv4 header. The type of service field consists of a 3-bit precedence subfield, a 4-bit TOS subfield, and the final bit which is unused and is set to be 0. The 4-bit TOS subfield enables 16 classes of service. In IPv6 header there is an 8-bit class of service field (see Figure 3.4). Later the Internet Engineering Task Force (IETF) differentiated services working group redefined IPv4 IPTOS to be DSCP, which is shown in Figure 3.4. DSCP has a 6-bit field enabling 64 classes of service.

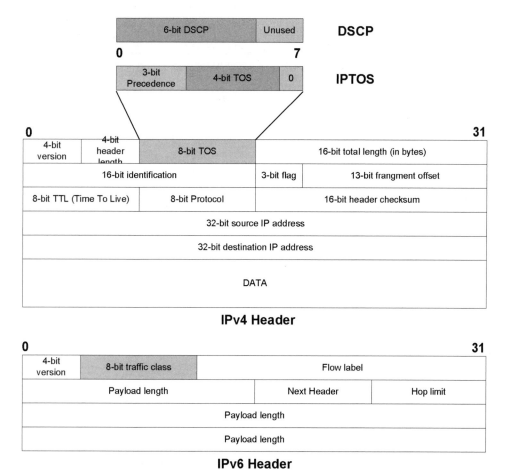

Figure 3.4 Structure of IPTOS and DSCP in IPv4 and IPv6

3.2.3 Transport Layer Classification (5-tuplet IP Header)

A 5-tuplet IP header (source IP, destination IP, source port, destination port, and protocol IP) can be used for transport layer classification. A 5-tuplet IP header can uniquely identify the individual application or flow. This classification provides the finest granularity and supports per-flow QoS service. However, the 5-tuplet IP header classification has some limitations:

- It is suitable for edge networks, but it is not suitable for core networks that carry very large amounts of traffic. Maintaining queues for each individual flow can be an overwhelming task.

• If the traffic passes through a firewall that uses NAT (network address translation), the real IP address (i.e., the IP address of the traffic source) is hidden from networks outside the firewall. Therefore, the 5-tuplet IP header exposed to a network outside the firewall cannot uniquely identify the application.

3.2.4 Application or User Classification

The application or user can be uniquely identified by using user/application identification (ID). The ID assignment may be static (i.e., the policy or the contract) or dynamic (i.e., connection signaling). For the connection signaling, there is a central station or entity in the network that is responsible for making the decision whether to allow a new session to join the network. First, the application or user sends the connection request to the central station. Then, if the new connection is admitted, it will be assigned a unique ID number. Packets from the application will be associated with an ID number.

3.3 Channel Access Mechanism

In wireless networks, all hosts communicate through a shared wireless medium. When multiple hosts try to transmit packets on the shared communication channel, collisions can occur. Therefore, wireless networks need a *channel access mechanism* which controls the access to the shared channel. There are two types of channel access mechanisms: 1) collision-based channel access and 2) collision-free channel access. Each type of channel access mechanism can provide different QoS services.

3.3.1 Collision-Based Channel Access

Collision-based channel access is a distributed channel access method that provides mechanisms to avoid collisions and to resolve collisions in case they occur. A classic collision-based channel access mechanism developed for wired LANs and implemented in Ethernet is CSMA/CD (Carrier Sense Multiple Access with Collision Detection). In collision-based channel access schemes, collisions can occur leading to the need for retransmissions. The collision probability depends on the number of active (with packets for transmission) users in the network. High traffic load increases the number of collisions and retransmissions, increasing the delay. Since we deal with stochastic traffic, the number of collisions and re-transmissions is random as well, leading to an unbound delay. Therefore, collision-based channel access schemes can provide best effort service. All hosts in the network receive equal bandwidth and experience the same unbounded delay. The service level can be improved by:

• Over-provisioning, whereby all traffic will receive ample of bandwidth and experience low delay.

- Adding a priority scheme in the collision-based channel access—that is, using different sized backoff windows for different priority classes. This will enable the provision of differentiated services. An example of such a solution is described in the proposed IEEE 802.11e (Chapter 4).

Existing solutions in wireless networks such as IEEE 802.11 DCF, HomeRF use collision-based channel access protocols similar to Ethernet CSMA/CD, denoted as CSMA/CA where CA stands for Collision Avoidance.

3.3.2   Collision-Free Channel Access

In a collision-free channel access mechanism the channel is arbitrated such that no collisions can occur. Only one host is allowed to transmit packets to the channel at any given time. Collision, therefore, will not occur. Examples of collision-free channel access techniques are polling and TDMA (Time Division Multiple Access).

3.3.2.1   Polling

A host in the network, or a specialized network device such as an Access Point or Base Station, is designated as the poller, which controls all access to the wireless channel by the other hosts denoted as pollees. Pollees are not allowed to transmit packets unless they receive a polling packet from the poller. Thus, there is no collision. Some pollees may receive the poll more often than others. The polling frequency (the number of polls in a period of time) reflects the bandwidth allocation. A poller can dynamically allocate bandwidth to pollees by adjusting the polling frequency dynamically.

3.3.2.2   TDMA (Time Division Multiple Access)

A TDMA scheme divides the channel access opportunity into frames and each frame is divided into time slots. A host is allowed to transmit packets in a predefined time slot, as shown in Figure 3.5.

The number of time slots assigned to a host per frame reflects the bandwidth allocated for the host. This technique requires a master host that is designated to manage the time slot assignment for all the hosts in the network. This Master host determines the number of time slots that each host will be allowed to transmit and notifies the hosts using some signaling mechanism. There are a number of time slot assignment philosophies:

- *Static time slot assignment:* Each host receives a fixed time slot assignment which can be provided during the connection setup.
- *Dynamic time slot assignment:* The time slot assignment changes dynamically during the lifetime of the session as a function of the traffic load, application QoS requirements, and channel conditions. This slot assignment policy is more flexible and leads to better channel utilization. However, it leads to signaling overhead required to communicate the slot assignment changes to the different hosts.

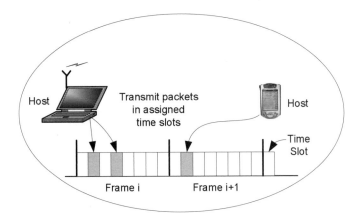

Figure 3.5 Time Division Multiple Access (TDMA) Scheme

Collision-free channel access schemes provide tight channel access control that can provide a tight delay bound. Therefore, these schemes are good candidates for QoS provision to applications with strict QoS requirements.

3.4 Packet Scheduling Mechanisms

Packet scheduling is the mechanism that selects a packet for transmission from the packets waiting in the transmission queue. It decides which packet from which queue and station are scheduled for transmission in a certain period of time. Packet scheduling controls bandwidth allocation to stations, classes, and applications.

As shown in Figure 3.6, there are two levels of packet scheduling mechanisms:

1. *Intrastation packet scheduling:* The packet scheduling mechanism that retrieves a packet from a queue within the same host.
2. *Interstation packet scheduling:* The packet scheduling mechanism that retrieves a packet from a queue from different hosts.

Packet scheduling can be implemented using hierarchical or flat approaches.

- *Hierarchical packet scheduling:* Bandwidth is allocated to stations—that is, each station is allowed to transmit at a certain period of time. The amount of bandwidth assigned to each station is controlled by interstation policy and module. When a station receives the opportunity to transmit, the intrastation packet scheduling module will decide which packets to transmit. This approach is scalable because interstation packet scheduling maintains the state by station (not by connection or application). Overall bandwidth is allocated based on stations (in fact, they can be groups, departments, or companies). Then, stations will have the authority to manage or allocate their own bandwidth portion to applications or classes within the host.

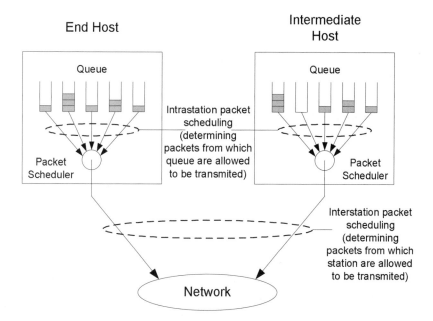

Figure 3.6 Packet Scheduling

- *Flat packet scheduling:* Packet scheduling is based on all queues of all stations. Each queue receives individual service from the network.

Packet scheduling mechanism deals with how to retrieve packets from queues, which is quite similar to a queuing mechanism. Since in intrastation packet scheduling the status of each queue in a station is known, the intrastation packet scheduling mechanism is virtually identical to a queuing mechanism. Interstation packet scheduling mechanism is slightly different from a queuing mechanism because queues are distributed among hosts and there is no central knowledge of the status of each queue. Therefore, some interstation packet scheduling mechanisms require a signaling procedure to coordinate the scheduling among hosts.

Because of the similarities between packet scheduling and queuing mechanisms we introduce a number of queuing schemes (First In First Out [FIFO], Strict Priority, and Weight Fair Queue [WFQ]) and briefly discuss how they support QoS services.

3.4.1   First In First Out (FIFO)

First In First Out (FIFO) is the simplest queuing mechanism. All packets are inserted to the tail of a single queue. Packets are scheduled in order of their arrival. Figure 3.7 shows FIFO packet scheduling.

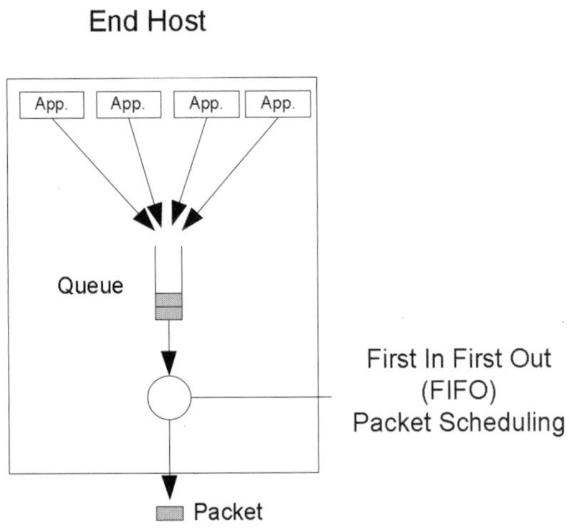

End Host

Figure 3.7   FIFO Packet Scheduling

FIFO provides best effort service—that is, it does not provide service differentiation in terms of bandwidth and delay. The high bandwidth flows will get a larger bandwidth portion than the low bandwidth flows. In general, all flows will experience the same average delay. If a flow increases its bandwidth aggressively, other flows will be affected by getting less bandwidth, causing increased average packet delay for all flows. It is possible to improve QoS support by adding 1) traffic policing to limit the rate of each flow and 2) admission control.

3.4.2   Strict Priority

Queues are assigned a priority order. Strict priority packet scheduling schedules packets based on the assigned priority order. Packets in higher priority queues always transmit before packets in lower priority queues. A lower priority queue has a chance to transmit packets only when there are no packets waiting in a higher priority queue. Figure 3.8 illustrates the strict priority packet scheduling mechanism.

Strict priority provides differentiated services (relative services) in both bandwidth and delay. The highest priority queue always receives bandwidth (up to the total bandwidth) and the lower priority queues receive the remaining bandwidth. Therefore, higher priority queues always experience lower delay than the lower priority queues. Aggressive bandwidth spending by the high priority queues can starve the low priority queues. Again, it is possible to improve the QoS support by adding 1) traffic policing to limit the rate of each flow and 2) admission control.

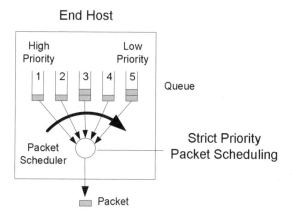

Figure 3.8 Strict Priority Packet Scheduling

3.4.3 Weight Fair Queue (WFQ)

Weight Fair Queue schedules packets based on the weight ratio of each queue. Weight, w_i, is assigned to each queue i according to the network policy. For example, there are three queues A, B, C with weights w_1, w_2, w_3, respectively. Queues A, B, and C receive the following ratios of available bandwidth: $w_1/(w_1+w_2+w_3)$, $w_2/(w_1+w_2+w_3)$, and $w_3/(w_1+w_2+w_3)$, respectively, as shown in Figure 3.9.

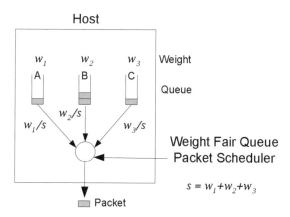

Figure 3.9 Weight Fair Queue Packet Scheduling

Bandwidth abuse from a specific queue will not affect other queues. WFQ can provide the required bandwidth and the delay performance is directly related to the allocated bandwidth. A queue with high bandwidth allocation (large weight) will experience lower delay. This may lead to some mismatch between the bandwidth and delay requirements. Some applications may require low bandwidth and low delay. In this case WFQ will allocate high bandwidth to these applications in order to guarantee the low delay bound. Some applications may require high bandwidth and high delay. WFQ still has to allocate high bandwidth in order for the applications to operate. Of course, applications will satisfy the delay but sometimes far beyond their needs. This mismatch can lead to low bandwidth utilization. However, in real life, WFQ mostly schedules packets that belong to aggregated flows, groups, and classes (instead of individual flows) where the goal is to provide link sharing among groups. In this case delay is of less concern.

The elementary queuing mechanisms introduced above will be the basis of a number of packet scheduling variations.

Before we move our discussion to the next QoS mechanisms, it is worth mentioning that in some implementations the channel access mechanism and packet scheduling mechanism are not mutually exclusive. There is some overlap between these two mechanisms and sometimes they are blended into one solution. When we discuss QoS support of each wireless technology in later chapters, in some cases, we will discuss both mechanisms together.

3.5 Traffic Policing Mechanism

Traffic policing is the mechanism that monitors the admitted sessions' traffic so that the sessions do not violate their QoS contract. The traffic policing mechanism makes sure that all traffic that passes through it will conform to agreed traffic parameters. When violation is found (e.g., more traffic is sent than was initially agreed upon in the QoS contract), a traffic policing mechanism is enforced by shaping the traffic. Because traffic policing shapes the traffic based on some known quantitative traffic parameters, multimedia (real-time) applications are naturally compatible to traffic policing. Most multimedia application traffic (voice, video) is generated by a standard codec which generally provides certain knowledge of the quantitative traffic parameters. Traffic policing can be applied to individual multimedia flows. Non-real-time traffic does not provide quantitative traffic parameters and usually demands bandwidth as much as possible. Therefore, traffic policing enforces non-real-time traffic (i.e., limits the bandwidth) based on the network policy. Such policing is usually enforced on aggregated non-real-time flows. Traffic policing can be implemented on end hosts or intermediate hosts. Examples of traffic policing mechanisms include the leaky bucket and the token bucket.

3.5.1 Leaky Bucket

The leaky bucket mechanism is usually used to smooth the burstiness of the traffic by limiting the traffic peak rate and the maximum burst size. This mechanism, as its name describes, uses the analogy of a leaky bucket to describe the traffic policing scheme. The bucket's parameters such as its size and the hole's size are analogous to the traffic policing parameters such as the

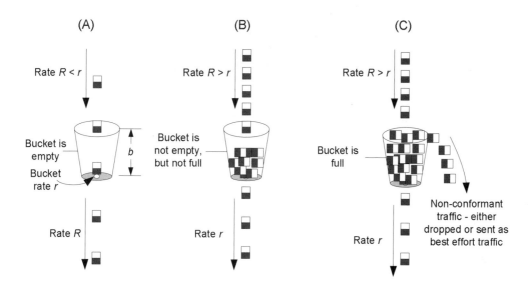

Figure 3.10 Leaky Bucket Mechanism

maximum burst size and maximum rate, respectively. The leaky bucket shapes the traffic with a maximum rate of up to the bucket rate. The bucket size determines the maximum burst size before the leaky bucket starts to drop packets.

The mechanism works in the following way. The arriving packets are inserted at the top of the bucket. At the bottom of the bucket, there is a hole through which traffic can leak out at a maximum rate of r bytes per second. The bucket size is b bytes (i.e., the bucket can hold at most b bytes). Let us follow the leaky bucket operation by observing the example shown in Figure 3.10. We assume first that the bucket is empty.

- Figure 3.10 (A): Incoming traffic with rate R which is less than the bucket rate r. The outgoing traffic rate is equal to R. In this case when we start with an empty bucket, the burstiness of the incoming traffic is the same as the burstiness of the outgoing traffic as long as $R < r$.
- Figure 3.10 (B): Incoming traffic with rate R which is greater than the bucket rate r. The outgoing traffic rate is equal to r (bucket rate).
- Figure 3.10 (C): Same as (B) but the bucket is full. Non-conformant traffic is either dropped or sent as best effort traffic.

3.5.2   Token Bucket

The token bucket mechanism is almost the same as the leaky bucket mechanism but it preserves the burstiness of the traffic. The token bucket of size b bytes is filled with tokens at rate r (bytes per second). When a packet arrives, it retrieves a token from the token bucket (given such a token is available) and the packet is sent to the outgoing traffic stream. As long as there are

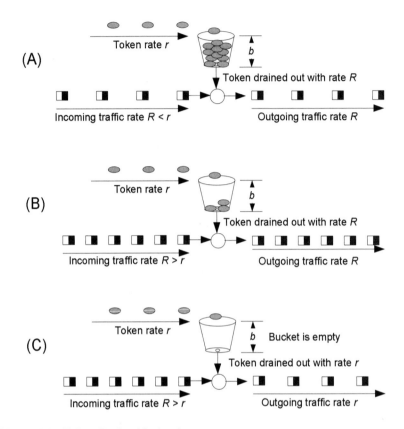

Figure 3.11 Token Bucket Mechanism

tokens in the token bucket, the outgoing traffic rate and pattern will be the same as the incoming traffic rate and pattern. If the token bucket is empty, incoming packets have to wait until there are tokens available in the bucket, and then they continue to send. Figure 3.11 shows an example of the token bucket mechanism.

- Figure 3.11 (A): The incoming traffic rate is less than the token arrival rate. In this case the outgoing traffic rate is equal to the incoming traffic rate.
- Figure 3.11 (B): The incoming traffic rate is greater than the token arrival rate. In case there are still tokens in the bucket, the outgoing traffic rate is equal to the incoming traffic rate.
- Figure 3.11 (C): If the incoming traffic rate is still greater than the token arrival rate (e.g., long traffic burst), eventually all the tokens will be exhausted. In this case the incoming traffic has to wait for the new tokens to arrive in order to be able to send out. Therefore, the outgoing traffic is limited at the token arrival rate.

The token bucket preserves the burstiness of the traffic up to the maximum burst size. The outgoing traffic will maintain a maximum average rate equal to the token rate, r. Therefore, the token bucket is used to control the average rate of the traffic.

In practical traffic policing, we use a combination of the token bucket and leaky bucket mechanisms connected in series (token bucket, then leaky bucket). The token bucket enforces the average data rate to be bound to token bucket rate while the leaky bucket (p) enforces the peak data rate to be bound to leaky bucket rate. Traffic policing, in cooperation with other QoS mechanisms, enables QoS support.

3.6 Resource Reservation Signaling Mechanisms

The traffic handling mechanisms (classification, channel access, packet scheduling, and traffic policing) we already described enable QoS services in each device. However, coordination between devices is essential to deliver end-to-end QoS services. Resource reservation signaling mechanisms inform the network entities on the QoS requirements of the multimedia applications using the network resources. The network devices will use this information to manage the network resources (i.e., bandwidth) in order to accommodate such requirements. The network devices control the network resources and provide QoS services by configuring the traffic handling mechanisms. Resource reservation can be applied to individual flows or aggregated flows. Resource reservation closely cooperates with the admission control mechanisms that will be described in a later section. Figure 3.12 shows a schematic diagram that describes the coordination between these mechanisms.

The resource reservation mechanisms include the following functions:

- Provision of resource reservation signaling that notifies all devices along the communication path on the multimedia applications' QoS requirements.
- Delivery of QoS requirements to the admission control mechanism that decides if there are available resources to meet the new request QoS requirements.
- Notification of the application regarding the admission result.

Resource Reservation Protocol (RSVP) is a well-known resource reservation signaling mechanism. RSVP operates on top of IP, in the transport layer, so it is compatible with the current TCP/IP based mechanisms (i.e., IPv4, IP routing protocol, and IP multicast mechanism) and can run across multiple networks. RSVP's main functionality is to exchange QoS requirement information among the source host, the destination host, and intermediate devices. Using this information, each network device will reserve the proper resources and configure its traffic handling mechanisms in order to provide the required QoS service. Once the reservation process is complete, the sender host is allowed to transmit data with an agreed traffic profile. If a device or network element on the communication path does not have enough resources to accommodate the traffic, the network element will notify the application that the network cannot support this QoS requirement. In order to achieve end-to-end resource reservation, all the network elements along

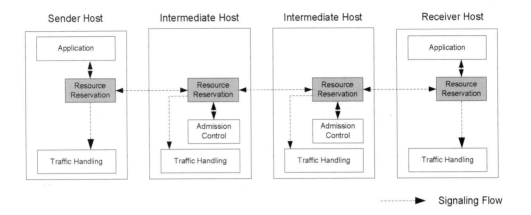

Figure 3.12 Resource Reservation Mechanism

the path (source host, destination host, and routers) need to be RSVP-enabled. Originally, RSVP was designed for supporting per-flow reservation. Currently it is extended to support per-aggregate reservation.

3.7 Admission Control

Admission control is the mechanism that makes the decision whether to allow a new session to join the network. This mechanism will ensure that existing sessions' QoS will not be degraded and the new session will be provided QoS support. If there are not enough network resources to accommodate the new sessions, the admission control mechanism may either reject the new session or admit the session while notifying the user that the network cannot provide the required QoS. Admission control and resource reservation signaling mechanisms closely cooperate with each other. Both are implemented in the same device. There are two admission control approaches:

- *Explicit admission control:* This approach is based on explicit resource reservation. Applications will send the request to join the network through the resource reservation signaling mechanism. The request that contains QoS parameters is forwarded to the admission control mechanism. The admission control mechanism decides to accept or reject the application based on the application's QoS requirements, available resources, performance criteria, and network policy.
- *Implicit admission control:* There is no explicit resource reservation signaling. The admission control mechanism relies on bandwidth over-provisioning and traffic control (i.e., traffic policing).

The location of the admission control mechanism depends on the network architecture. For example, in case we have a wide area network such as a high-speed backbone that consists of a number of interconnected routers, the admission control mechanism is implemented on each router. In shared media networks, such as wireless networks, there is a designated entity in the network (e.g., station, access point, gateway, base station) that hosts the admission control agent. This agent is in charge of making admission control decisions for the entire wireless network. This concept is similar to the SBM (subnet bandwidth manager) which serves as the admission control agent in 802 networks.

In ad hoc wireless networks, the admission control functionality can be distributed among all hosts. In infrastructure wireless networks where all communication passes through the access point or base station, the admission control functionality can be implemented in the access point or base station.

3.8 QoS Architecture

This section shows how all the QoS mechanisms described in the previous subsections are working together to provide QoS support. Different applications that co-exist in the same network may require different combinations of QoS mechanisms such as:

- *Applications with quantitative QoS requirements:* These applications mostly require QoS guaranteed services. Therefore, explicit resource reservation and admission control are needed. They also require strict traffic control (traffic policing, packet scheduling, and channel access).
- *Applications with qualitative QoS requirements:* These applications require high QoS levels but do not provide quantitative QoS requirements. In this case we can use resource reservation and admission control. They also require traffic handling which delivers differentiated services.
- *Best effort:* There is no need for QoS guarantees. The network should reserve bandwidth for such services. The amount of reserved bandwidth for best effort traffic is determined by the network policy.

The QoS architecture which contains different QoS mechanisms is different for each network topology. We will focus on the QoS architecture for ad hoc and infrastructure wireless networks.

3.8.1   QoS Architecture for Infrastructure Wireless Networks

In infrastructure wireless networks, there are two types of stations: end stations (hosts) and a central station (i.e., access point, base station). The central station regulates all the communication in the network—that is, there is no peer-to-peer communication that occurs directly between the hosts. The traffic from a source host is sent to the central station and then the central station forwards the traffic to the destination host. All traffic handling (classification, traffic policing, packet scheduling, and channel access) and resource reservation mechanisms reside in all stations (end hosts and central station). In addition, the central station also includes an admission control mechanism. Figure 3.13 shows a QoS architecture for an infrastructure wireless network.

There are some variations in the signaling mechanisms that configure the traffic handling mechanisms in each station. We will point out these differences in each wireless technology chapter.

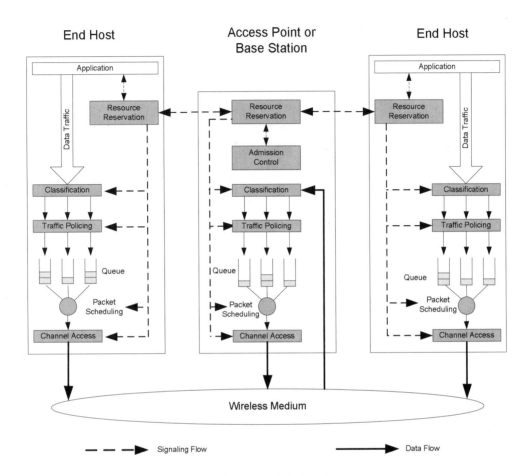

Figure 3.13 QoS Architecture of an Infrastructure Wireless Network

3.8.2   QoS Architecture For Ad Hoc Wireless Networks

All hosts establish peer-to-peer communication in the shared wireless media environment. All traffic handling and resource reservation mechanisms reside in all hosts. One of the hosts (either a dedicated or a regular end host) will be designated to serve as an admission control agent (i.e., designated SBM [DSBM]). Figure 3.14 shows a QoS architecture for an ad hoc wireless network.

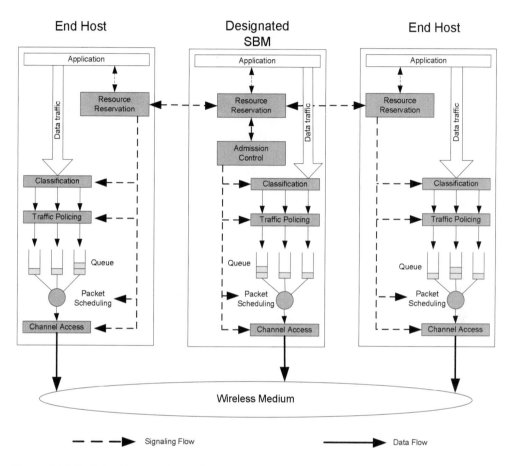

Figure 3.14 QoS Architecture for an Ad Hoc Wireless Network

Wireless Local Area Networks

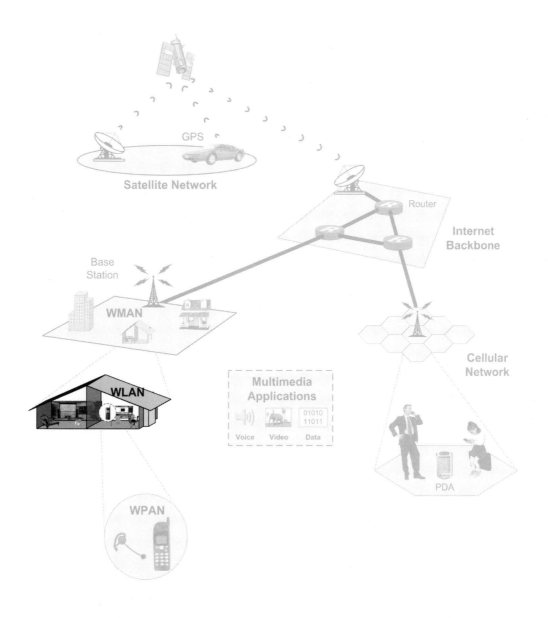

CHAPTER 4

IEEE 802.11

4.1 IEEE 802.11

As shown in Figure 4.1, the IEEE (Institute of Electrical and Electronics Engineers) standards committee has defined a family of IEEE 802 LAN (local area network) and MAN (metropolitan area network) standards in which IEEE 802.11 defines the wireless LAN (WLAN) standard with multiple physical layer options including speeds of up to 54 Mbps. The organization is expected to introduce enhancements to the standard that will support QoS and other radio options. Many companies (e.g., Cisco, 3Com, Linksys, D-Link) have introduced communication cards based on this standard.

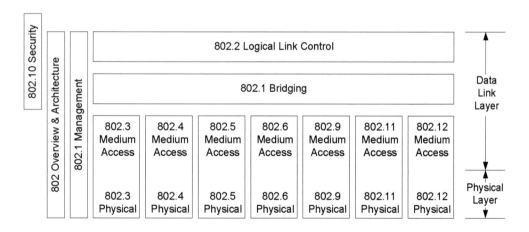

Figure 4.1   IEEE 802 Standards

Compatibility and interoperability among various IEEE 802.11 vendors are managed by the Wireless Ethernet Compatibility Alliance (WECA). Their certification is referred as Wi-Fi, which stands for Wireless Fidelity. Wi-Fi certification gives consumers and businesses the assurance that WLAN products bearing the Wi-Fi logo are compatible and work together even when manufactured by different vendors. Such PC products include PCMCIA (PC Memory Card International Association) cards for notebooks, PCI (peripheral component interconnect/interface) cards for desktops, and cards with USB (Universal Serial Bus) interface that can be used with either one.

The wireless media are fundamentally different from the wired media in the following aspects:

- The wireless media have no clear physical boundaries in which users can transmit and receive data.
- Users can sometimes be covered by multiple WLANs. This fact requires a resolution mechanism that will enable the users to conduct effective communication with minimal interruptions from other WLANs' participants.
- Users are susceptive to radio interferences coming from sources such as microwaves, causing the wireless media to be significantly less reliable then the wired media.
- Users are mobile—that is, they can change their locations, and roam within their coverage areas. This mobility causes a dynamic network topology, which leads to the need to have procedures for the users to associate/disassociate with the specific WLAN. Moreover, a user can be temporarily hidden from some other network members.
- The power limitation, the wireless media nature, and the topographic characteristics of the communication area impose a limited range for the effective geographical coverage. This fact may require the installation of several overlapping wireless LANs to cover a larger geographical area.
- Many, if not most, of the wireless devices are battery powered. Hence, power saving is essential to increase the time between battery charges. This fact may cause situations in which the user is temporarily without service.

The IEEE 802.11 family consists of a few standards. The first standard that was approved in 1997 details the Wireless LAN Medium Access Control (MAC) and Physical Layer (PHY) specifications that support data rates of 1 Mbps and 2 Mbps over the 2.4 GHz Industrial, Scientific, and Medical (ISM) frequency band using either Frequency Hopping Spread Spectrum (FHSS) or Direct Sequence Spread Spectrum (DSSS) as their radio technologies as well as infrared. (Note: ISM frequency bands are nonlicensed frequency bands used for wireless network, cordless phone, and other devices. ISM frequency bands include 902–928 MHz, 2.4–2.5 GHz and 5.725–5.875 GHz.) The standard details operational specifications for wireless connectivity for fixed, ad hoc, and mobile stations within a local geographical area. This first standard was introduced after years of careful development and has become popular with vendors who implemented significant parts of this standard.

The IEEE 802.11 standard contains two additions introduced later: IEEE 802.11a enhances the speed to up to 54 Mbps using Orthogonal Frequency Division Multiplexing (OFDM) radio technology in the 5 GHz band, and IEEE 802.11b enhances the speed to 5.5 Mbps and 11 Mbps using Complementary Code Keying (CCK) modulation in the 2.4 GHz band. These standards allow for multirate support and backwards compatibility (i.e., IEEE 802.11b supports speeds of 1, 2, 5.5, and 11 Mbps). As long as users in the same WLAN use the same basic radio technology, even though with different data rates, these users will be able to communicate using the lowest speed. For example, a lower speed DSSS based radio (i.e., IEEE 802.11) will be able to communicate with a more advanced CCK based radio (i.e., IEEE 802.11b) sharing the lower speed (1 or 2 Mbps) of the two radios. The IEEE 802.11g committee is working on the definition of a standard in which higher speeds of more than 20 Mbps will be used within the 2.4 GHz band. This potential extension will be backwards compatible and interoperable with IEEE 802.11b. IEEE 802.11g provides speeds of 1, 2, 5.5, 6, 9, 11, 12, 18, 24, 36, 48, and 54 Mbps.

The IEEE 802.11 standards consider data traffic but lack proper consideration for multimedia applications' QoS needs. Currently, the IEEE 802.11 group is working on an extension, IEEE 802.11e, that will specify how signaling and support for QoS can be achieved.

Furthermore, the IEEE 802.11 group is also working on a number of standard supplements as summarized in Table 4.1.

Table 4.1   IEEE 802.11 Standard Supplements

Standard Supplements	Brief Description
IEEE 802.11a	Define a PHY in newly allocated UNII band. The standard has been completed and published as 8802-11: 1999 (E)/Amd 1: 2000 (ISO/IEC) (IEEE Std. 802.11a-1999 Edition).
IEEE 802.11b	Define a high rate PHY in the 2.4 GHz band. The standard has been completed and published as IEEE Std. 802.11b-1999.
IEEE 802.11c	Define MAC procedure for the bridge operation. The standard has been completed and is now part of IEEE 802.11c standard.
IEEE 802.11d	Define PHY requirements to extend the operation of 802.11 WLAN to new regulatory domains (countries). The standard has been completed and published as IEEE Std. 802.11d-2001.
IEEE 802.11e	Enhance current 802.11 MAC to expand support for applications with QoS requirement.

Table 4.1   IEEE 802.11 Standard Supplements (Continued)

Standard Supplements	Brief Description
IEEE 802.11f	Recommend practice for Inter Access Point Protocol (IAPP).
IEEE 802.11g	Standard for high-speed (20+ Mbps) PHY extensions to the 802.11b standard.
IEEE 802.11h	Enhance 802.11 MAC and 802.11a PHY with network management, control extensions for spectrum and transmit power management in 5GHz license exempt bands.
IEEE 802.11i	Enhance the 802.11 MAC to enhance security and authentication mechanisms.
IEEE 802.11j	Enhance the current 802.11 MAC and 802.11a PHY to additionally operate in newly available Japanese 4.9 GHz and 5 GHz bands.
IEEE 802.11k	Define Radio Resource Measurement enhancements to provide interfaces to higher layers for radio and network measurements.

IEEE 802.11a, b, c, and d have already been approved, whereas IEEE 802.11e, f, g, h, i, j, and k are under development.

4.1.1   Architecture

4.1.1.1   Network Topologies

The standard addresses two topologies: ad hoc topology, referred to as Independent Basic Service Set (IBSS), and infrastructure topology, referred to as Basic Service Set (BSS). A topology that combines several BSS cells is referred to as Extended Service Set (ESS).

4.1.1.1.1 Ad Hoc Network   As shown in Figure 4.2, an ad hoc network or an IBSS consists of stations within mutual communication range of each other via the wireless medium. Such a network is created spontaneously, without preplanning, for ad hoc temporary situations with limited needs to access the Internet. The IBSS is the most basic type of an IEEE 802.11 WLAN and may contain only two stations. Figure 4.2 shows two IBSSs, each with two stations. If a station moves out of its IBSS, meaning out of range, it can no longer communicate with the other IBSS members.

Figure 4.2   Ad Hoc Topology: Independent Basic Service Sets

4.1.1.1.2 Infrastructure Network   The infrastructure network or BSS includes an access point (AP) in addition to the stations. This AP acts as the BSS arbitrator, meaning that the AP will handle all the BSS traffic. The BSS traffic can be either internal traffic (i.e., among the BSS participants) or external traffic (between the BSS participants and outside the BSS). The AP integrates the BSS within the distribution network. For example, all traffic between the BSS participants and the Internet will be delivered through the AP.

Figure 4.3 shows an ESS composed of two BSSs, each with two stations and an AP. Each BSS is interconnected to the distribution system (DS) (which may connect to the Internet) through the AP. The key concept is that the ESS network appears as a single network entity to the upper layers including the applications.

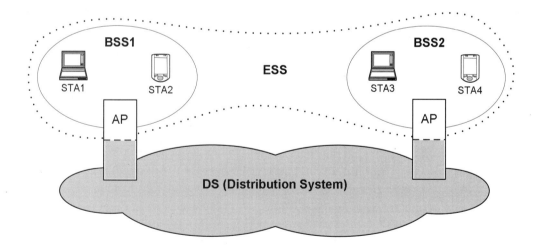

Figure 4.3   Infrastructure Topology: Extended Service Set

To allow effective communication for users who constantly move from one BSS geographical area to another BSS area, the BSSs should be physically overlapping. In the standard there is no limit to the distance between the BSSs and it is left up to the user or network installer to determine the BSS location. Moreover, other users may decide to install an IBSS in the same geographical area. In this case, multiple WLANs will be collocated and function effectively if the respective users coordinate the used frequencies.

4.1.1.1.3 Collocation   Wireless networks can be located in overlapping geographical areas (see Figure 4.4). Radio signals may propagate from one network to another. A change in the transmitter position and geographical layout may have a profound effect on the propagation characteristics. Hence such networks do not have a clear boundary. The possibility of having multiple collocated operational WLANs not only allows multiple independent user groups to work simultaneously in the same place but also aggregates and increases the WLAN capacity.

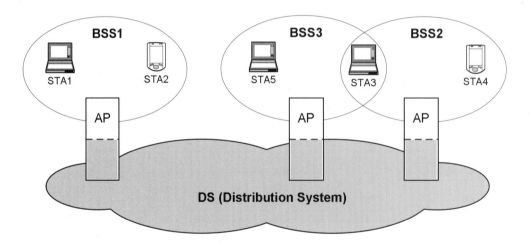

Figure 4.4   Collocated WLANs

Collocation is possible if network installers make an effort to coordinate the frequencies used by their respective WLANs. If they use DSSS based WLANs they can coordinate the center frequency. Several centers are possible according to the region in which the networks are installed. For example, IEEE 802.11b in North America details three non-overlapping channels and six overlapping channels (see Figure 4.5). This means that up to three networks can be located without mutual interruptions.

Collocation is also possible in FHSS-based WLANs. IEEE 802.11b defines 78 hopping sequences (each with 79 hops) grouped in three sets of 26 sequences each. This means that theoretically 26 FHSS WLAN systems can be collocated. However, practically fewer user groups can be collocated because of interference from other groups. This number is estimated to be 15. In IEEE 802.11a collocation is possible if different WLANs use different operating frequencies.

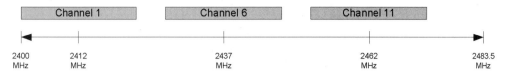

North American Channel Selection - non-overlapping (2.4 GHz ISM frequency band)

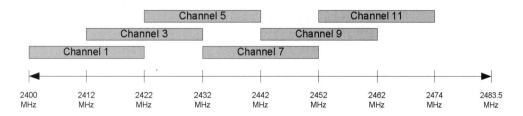

North American Channel Selection - overlapping (2.4 GHz ISM frequency band)

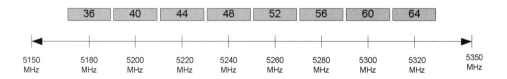

North American Channel Selection (Lower U-NII 5150 - 5250 MHz, Middle U-NII 5250 - 5350 MHz)

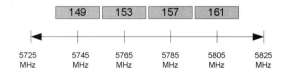

North American Channel Selection (Upper U-NII 5725 - 5825 MHz)

Figure 4.5 North American Channels

In the U.S., there are three 100 MHz unlicensed national information infrastructure (U-NII) bands: Lower U-NII (5.15 to 5.25 GHz), Middle U-NII (5.25 to 5.35 GHz), and Upper U-NII (5.725 to 5.825 GHz).

4.1.1.2   Protocol Stack

As shown in Figure 4.6, the IEEE 802.11-1997 base standard focuses on the following two parts of the protocol stack: the MAC part of the data link layer and the physical layer.

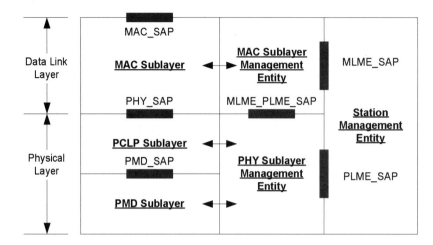

Figure 4.6   Portions of the OSI Protocol Stack Covered by IEEE 802.11 Base Standard
(PLCP [physical layer convergence protocol]). Abbreviations: MLME—MAC
Sublayer Management Entity; PLME—Physical Layer Management Entity;
SAP—Service Access Point.

4.1.2   Physical Layer

IEEE 802.11 allows for various wireless technologies: infrared and radio DSSS, FHSS, and
OFDM. One technology cannot work with the other. In other words, radio FHSS cannot communicate with a radio that employs DSSS. The standard, though, specifies backwards compatibility
for radios that use the same transmission technology. In other words, a higher speed radio will be
able to communicate with a lower speed radio via adjusting the speed of the higher speed radio
to that of the lower speed radio. IEEE 802.11a allows for higher speeds in the 5 GHz band based
on OFDM: 6, 9, 12, 18, 24, 36, 48, and 54 Mbps (support of 6, 12, and 24 Mbps data rates is
mandatory). OFDM is also considered by IEEE 802.11g at the 2.4 GHz frequency range to
achieve speeds of 1, 2, 5.5, 6, 9, 11, 12, 18, 24, 36, 48, and 54 Mbps.

The various radio flavors use slightly different frequencies in different countries and continents. This means that a U.S. radio will not work properly with a Japanese radio. Ongoing work
is conducted to allow better interoperability and to minimize the need for equipment manufacturers to produce a wide variety of country-specific products and for users to travel with a bag
full of country-specific WLAN cards. Such effort is reflected in the IEEE 802.11d that supplements the IEEE 802.11 with features that allow WLANs to operate within the rules of different
countries. Another effort is IEEE 802.11f, whose goal is to achieve radio access point interoperability within a multivendor WLAN environment.

4.1.3   Media Access Control (MAC)

The MAC governs the stations' access to the shared wireless medium. The MAC which is located with the data link layer has a crucial role in providing QoS support to users, especially when executing multimedia applications.

The MAC architecture details two operational modes that coexist (see Figure 4.7): the Distributed Coordination Function (DCF) and the Point Coordination Function (PCF). The two modes use a cycle structure (see Figure 4.8) denoted collision-free period repetition interval, in which the first time period (called Contention-free Period or CFP) is governed by the PCF mode and the second time period (called Contention Period or CP) is governed by the DCF mode. DCF is a simplistic mode that allows contention and collision of traffic between stations. Therefore, DCF is suitable for applications that do not require QoS. PCF is a more complex protocol that uses the Point Coordinator (PC) which resides at the AP. The PC arbitrates the channel access using a polling based approach. PCF is geared to provide QoS support for multimedia applications. However, there are no known products that implement the PCF mode. QoS provisioning is going through an extensive review by the IEEE 802.11 organization under the extension IEEE 802.11e. This is discussed in Section 2 of this chapter.

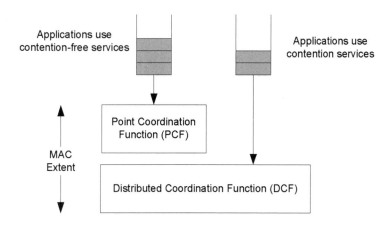

Figure 4.7   IEEE 802.11 MAC Architecture

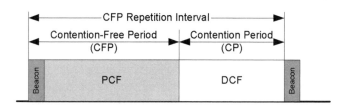

Figure 4.8   IEEE 802.11 Collision-Free Period Repetition Interval

In addition to defining the media access rules for each mode, the standard also defines procedures for traffic fragmentation and defragmentation, multirate support, authentication and privacy. Fragmentation, or slicing the original data frame into smaller transmitted frames, is carried in order to reduce the loss of traffic which can occur due to collisions with other users' traffic or due to radio interference. The smaller the frame, the less wireless media bandwidth is lost in the collision. However, smaller frames lead to more overhead consisting of frames headers that contain addresses and other pertinent information. Hence, there is a tradeoff which will determine how much fragmentation should be carried out. The standard does not quantify the level of fragmentation and leaves it to the implementers' discretion.

4.1.3.1   Distributed Coordination Function (DCF)

The fundamental channel access mechanism of the IEEE 802.11 MAC is DCF, also known as Carrier Sense Multiple Access with Collision Avoidance (CSMA/CA). This is a random access mechanism that enables sharing of the wireless media between compatible radio transmitters. The standard mandates that the DCF is implemented in all stations for use within both the IBSS and BSS configurations.

In the DCF mode a station cannot transmit arbitrarily, but rather needs to go through a set of steps to determine whether it can transmit. The goal of these steps is to reduce the chance of collision with other stations' packets. Such collision causes the destruction of the transmitted packet along with other stations' packets. Before transmission, the user's transmitter senses the wireless medium to determine if another station is transmitting. If the wireless medium is determined to be free, the transmitter will sense the channel and wait before transmission for a certain time referred to as Interframe Space (IFS). Once the data are successfully received by the addressee, the addressee is required to respond with an immediate positive acknowledgment (ACK) such that the user is assured that the data have arrived. In case the ACK is not received, the user station will go through a process of retransmission until an ACK is received. When a retransmission is required, the transmitter needs to wait an additional random backoff time before transmitting in order to minimize collisions with the other stations that also participated in the collision.

IEEE 802.11 defines four IFSs with different duration (see Figure 4.9): Short Interframe Space (SIFS), PCF Interframe Space (PIFS), DCF Interframe Space (DIFS), and Extended Interframe Space (EIFS). The duration of these intervals is determined in the standard as a function of the radio technology used. The duration of these intervals determines the priority of the packets (i.e., in case the packet waits SIFS, which is shortest interframe interval, the packet has the highest priority).

The IFS intervals are used as follows:

- SIFS is used as an interframe space between a data frame and its ACK frame, between an RTS (request to send) frame and its CTS (clear to send) frame between fragments of the original data frame, and by a station responding to a control message in the PCF mode.

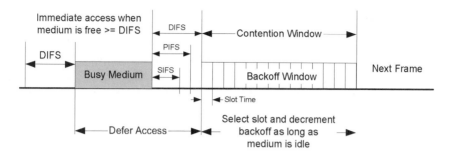

Figure 4.9 IEEE 802.11 Interframe Spacing (IFS)

- PIFS is used in the PCF mode to gain priority access to the medium.
- DIFS is used in the DCF mode to transmit data and management frames. If the wireless medium is free for the DIFS period the station is allowed to transmit.

Despite this elaborate mechanism collisions can occur if stations are far away from each other and cannot listen to each other's transmissions or if the stations start transmitting concurrently. The more stations there are in the BSS or IBSS the more collisions occur, and hence bandwidth utilization is deteriorated.

The standard provides a mechanism to further minimize collisions by reserving the channel for a certain period of time. During this period of time only one station is allowed to transmit while others defer. This reservation is obtained through the following process. The transmitting station (i.e., the station that would like to reserve the channel) transmits a special announcement to all stations, announcing its intent to transmit in the period mentioned in this announcement. The addressed station is expected to respond and acknowledge this announcement. The announcement is referred to as Request To Send (RTS) packet and the addressee's response is referred to as Clear To Send (CTS) packet. All stations who receive either the RTS or CTS will refrain from their transmission attempts until the end of the time period announced in either the RTS or CTS packets. The RTS/CTS mechanism cannot be used for broadcast and multicast traffic. The RTS/CTS burdens the network with additional traffic and overhead that reduces some of its benefit. The standard does not define when and in what situations the RTS and CTS exchange should be carried out. This decision is left to the implementer and even the user in cases where the wireless communication card provides access to this decision.

Based on the CSMA/CA protocol, each station maintains a prediction of time in which future traffic will occur. This prediction mechanism is referred in the standard as the Network Allocation Vector (NAV). This duration is announced either in the RTS/CTS packets or in the header of the packets sent during the DCF mode. The stations use the NAV to defer from accessing the wireless medium.

4.1.3.2   Point Coordination Function (PCF)

The IEEE 802.11 MAC standard defined an optional access method called Point Coordination Function (PCF) which operates during the Contention-Free Period (CFP). This mode can be used only in an infrastructure network and uses the Point Coordinator (PC) as the network coordinator or arbiter. In a practical implementation, the PC can be included in the AP. Using a centralized polling based approach, the PC determines which station has the right to transmit and polls this station. A station can transmit only when polled by the PC. When polled, the station can transmit only one data frame which can be either a new data packet or a retransmission packet of a previous packet for which no ACK has been received. The station is allowed, though, to piggyback an ACK on the transmitted data frame. Thus, the PCF provides a contention-free data transfer.

To gain access to the medium and start the PCF period (in CFP), the PCF mode uses the PIFS interval (which is shorter in duration than the DIFS interval) along with special management frames that include the duration of the network allocation vector (NAV) to be set by all other stations. Using PIFS, the PC has the highest channel access priority. The PC sends the NAV value in the Beacon message. All stations set their NAV values which prevent them from accessing the channel during the CFP, therefore preventing collisions.

Thus, PCF provides collision-free access to the channel which has the potential to provide QoS support. As discussed later in this chapter, the standard does not provide the polling intelligence in terms of when to poll each station. It is left for the implementer to decide the polling intelligence.

4.1.4   Physical Layer Convergence Protocol (PLCP)

The PLCP layer located beneath the MAC has several responsibilities. It carries out the Carrier Sensing (CS) of the wireless media to determine whether there are ongoing transmissions on the channel. It also synchronizes the speed of the communicating stations' transmitter and receiver and establishes the communication. To accomplish these responsibilities it adds additional information to each packet.

The CS mechanism can be configured with several Clear Channel Assessment (CCA) options:

- *CCA Mode 1:* Energy above threshold. CCA reports a busy medium upon detecting any energy above the threshold.
- *CCA Mode 4:* Carrier sense with timer. CCA starts a timer whose duration is defined in the standard and reports a busy medium only upon the detection of a signal until the timer expires. Otherwise it reports the media is not busy.
- *CCA Mode 5:* A combination of carrier sense and energy above threshold.

The synchronization is obtained by using packet headers which are transmitted in increasing speeds. This means that the first part of the header is transmitted in a low speed, the second in a higher speed, and so forth. This mechanism allows speed synchronization but also slows down higher speed radios by as much as 30%.

4.1.5 QoS Support

As we mentioned previously, IEEE 802.11 MAC has two modes of operations, PCF and DCF, which deliver different QoS support. In order to clearly understand the QoS support, we first examine the QoS mechanisms provided by IEEE 802.11.

4.1.5.1   QoS in IEEE 802.11 DCF

QoS mechanisms (Figure 4.10)

1. *Classification:* There is no classification in DCF. Therefore, no service differentiation is provided.
2. *Channel access:* DCF uses contention-based media access control protocol (CSMA/CA).
3. *Packet scheduling:* The intrastation packet scheduling uses FIFO mechanism (i.e., all traffic in a station is queued and transmitted in a first-in-first-out order).

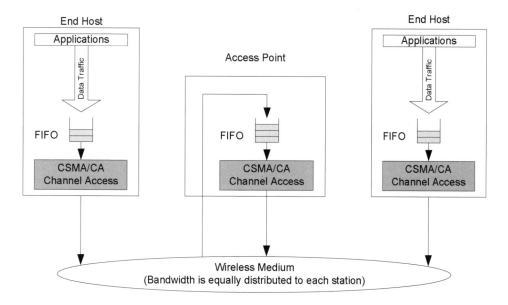

Figure 4.10 IEEE 802.11 DCF QoS Mechanisms

QoS service and supported applications
DCF delivers best effort QoS service level which is suitable for non-real time applications. There is no service differentiation and no service guarantee in terms of bandwidth and delay.

4.1.5.2   QoS in IEEE 802.11 PCF

QoS Mechanisms (Figure 4.11)

    1. *Classification:* There is no classification in PCF. Therefore, there is no service differentiation within stations (i.e., all traffic in the same station is treated equally).

    2. *Channel access:* PCF uses a polling based media access control protocol. A station is allowed to transmit packets only when it receives a polling message from the PC. Since no collisions occur, PCF can deliver a predictable service performance.

    3. *Packet scheduling:* Due to the lack of a classification mechanism within each station, the intrastation packet scheduling uses a FIFO mechanism. The interstation packet scheduling is directly related to the polling sequence and polling frequency (the number of polls to a station in a time period) which are not specified by the standard.

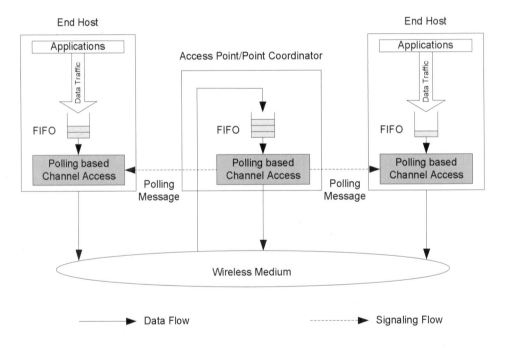

Figure 4.11 IEEE 802.11 PCF QoS Mechanisms

QoS service and supported applications

PCF can deliver a certain level of guaranteed QoS service which is suitable for real-time applications. To achieve guaranteed QoS service, the following functions have to be explicitly determined:

- Algorithms to determine the polling sequence and polling frequency to properly allocate bandwidth and satisfy applications' delay requirements.
- Algorithms to determine the duration of the CFP repetition interval, CFP interval, and CP interval so that applications' delay requirements are satisfied.
- QoS signaling: to determine the proper polling sequence and polling frequency, the PC needs to obtain from the stations information on the traffic dynamics at each station as well as applications' quantitative QoS requirements. Certain QoS signaling is included in IEEE 802.11e.

4.2 IEEE 802.11e (QoS Extension)

Realizing the shortcomings of the QoS support within the current IEEE 802.11 MAC, the IEEE organization (IEEE 802.11 Task Group E) has been involved in an extensive effort to specify MAC enhancements for better QoS support. This effort and potential extension to the standard are referred to as IEEE 802.11e. This extension has not been approved yet. Hence this discussion provides just a simplified explanation and glimpse into the proposals and ideas discussed within this effort. The proposed IEEE 802.11e applies to the physical layers defined in IEEE 802.11 and IEEE 802.11a, b, and g. In addition, the proposal allows for QoS support even when legacy IEEE 802.11 stations are present. This is an important feature that allows for friendly coexistence of new IEEE 802.11e compliant stations with legacy IEEE 802.11 stations.

Several problems with the PCF motivated this effort. Such problems include the unpredictable delays associated with beacons due to the CSMA/CA and PIFS mechanisms, unknown transmission durations of the polled stations caused by unknown frame lengths and modulation techniques, and the inability to differentiate among traffic streams which can enable the provision of higher priorities to QoS sensitive applications. These problems soften the QoS support in the following ways: 1) time delay is unbounded causing delay sensitive transmissions to be queued for a lengthy time, and 2) there is an inability to provide the needed bandwidth guarantees for bandwidth sensitive applications.

IEEE 802.11e's goal is to deliver the following:

- Quantitative QoS services (parameterized QoS): It provides QoS services that meet the application's quantitative requirements in terms of traffic specifications (TSPEC).
- Differentiated services (prioritized QoS): It provides priority services among traffic streams.

IEEE 802.11e proposes the following QoS enhancements:

- Traffic classification: traffic category and traffic stream
- Channel access and packet scheduling: enhanced MAC functions such as Enhanced DCF (EDCF) and Hybrid Coordination Function (HCF)
- QoS signaling
- New frames for QoS support

4.2.1 Traffic Classification

802.11e provides two types of classification: traffic category (TC) and traffic stream (TS).

4.2.1.1 Traffic Category

A traffic category provides a tool for applications to set a distinct priority relative to other data to be transmitted over the wireless media (prioritized QoS). 802.11e defines traffic categories to support differentiated services to at most eight delivery priorities designated 0 through 7. Data streams are classified to one of the eight traffic categories. By default, user priority TC7 is mapped to the highest delivery priority and user priority TC1 is mapped to the lowest delivery priority (see Table 4.2). It is possible to have multiple data streams mapped to the same traffic category. Therefore, classification based on traffic category is comparable to per-class classification defined in Chapter 3. Traffic Category cooperates with EDCF.

Table 4.2   IEEE 802.11e Traffic Categories

Priority Level	User Priority	Traffic Categories	Acronym	Traffic Type
Lowest	1	TC1	BK	Background
	2	TC2	--	Spare
	0 (default)	TC0	BE	Best effort
	3	TC3	EE	Excellent effort
	4	TC4	CL	Controlled load
	5	TC5	VI	Video <100 msec delay and jitter
	6	TC6	VO	Voice < 10msec delay and jitter
Highest	7	TC7	NC	Network control

To manage these traffic categories, IEEE 802.11e has to implement a number of physical queues. Table 4.3 shows a mapping between the traffic categories and the queues.

Table 4.3   IEEE 802.11e Mapping of Traffic Category to Queues

Number of Queues	Defining Traffic Type							
1	BE							
2	BE				VO			
3	BE			CL		VO		
4	BK		BE	CL		VO		
5	BK		BE	CL	VI	VO		
6	BK	BE	EE	CL	VI	VO		
7	BK	BE	EE	CL	VI	VO	NC	
8	BK	--	BE	EE	CL	VI	VO	NC

4.2.1.2   Traffic Stream

To support quantitative QoS services (parameterized QoS), IEEE 802.11e defines traffic streams (TSs). In each station there are at most eight traffic streams. Each traffic stream is associated with its traffic specification (TSPEC) which includes:

1. Quantitative objectives for traffic attributes such as packet size and arrival rates
2. Traffic characteristics (constant vs. variable data rate, maximum delivery delay, maximum delay variance, etc.)
3. Acknowledgment policy

Each data flow with quantitative QoS requirement is directly classified to a traffic stream. Classification based on traffic streams is comparable to per-flow classification defined in Chapter 3. Traffic category cooperates with HCF.

Figure 4.12 shows the classification (both traffic category and traffic stream) in a station.

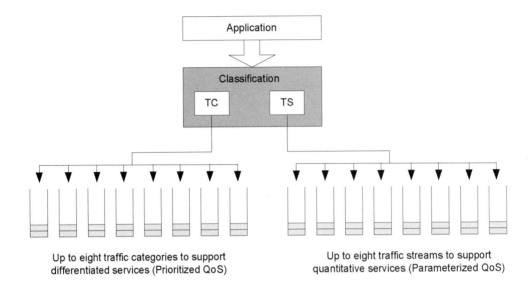

Figure 4.12 IEEE 802.11e Classification Mechanisms

4.2.2   **Channel Access and Packet Scheduling**

802.11e provides enhanced MAC functions (EDCF, HCF) which include both channel access and packet scheduling mechanisms. Therefore, we discuss both QoS mechanisms at once for each MAC.

    IEEE 802.11e defines the following MAC protocol with two modes of operation: the Hybrid Coordination Function (HCF) to support QoS network configurations and Enhanced DCF (EDCF). Similar to 802.11, 802.11e proposal has two phases of MAC operation: Contention Period (CP) and Contention-Free Period (CFP), which alternate over time continuously. In each phase the station can send data packets or their fragments, according to rules set by each phase. The EDCF is used in the CP phase only, while the HCF is used in both phases (see Figure 4.13).

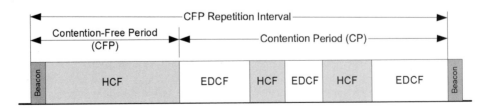

Figure 4.13 IEEE802.11e CFP Repetition Interval

The HCF combines functions from the legacy DCF and PCF with some enhanced, QoS-specific functions and frame subtypes. Such enhancements allow a uniform set of frame exchange sequences to be used for QoS transfers during both the CP and CFP. The HCF uses a contention based media access method, referred as enhanced DCF (EDCF). Stations may obtain transmission opportunities using one or both of these media access methods. Stations that support the 802.11e are referred as QoS Stations (QSTAs). The QoS station has an option to function as the centralized controller for all other stations within the WLAN or QoS supporting Basic Service Set (QBSS). This centralized controller is called Hybrid Coordinator (HC) or Point Coordinator (similar to 802.11 PCF). The HC is expected to be installed in the 802.11e Access Point (AP).

4.2.2.1   Enhanced Distributed Coordination Function (EDCF)

The EDCF provides differentiated, distributed access to the wireless medium for at most eight delivery priorities as defined above. EDCF channel access is derived from CSMA/CA used in DCF with the addition of priorities. Multiple packets are delivered through multiple backoff instances within one station. Each backoff instance is labeled with a TC specific parameter. In the CP phase, each TC within the stations contends for transmission and independently starts a backoff process congruent to the CSMA/CA after detecting the channel being idle for an Arbitration Interframe Space (AIFS). The AIFS is at least DIFS, and can be increased for each TC. Consequently, data streams with higher AIFS values, meaning longer backoff times, have lower priority access to the wireless media. Data streams with lower AIFS value will be able to access the wireless media before data streams with higher AIFS values (see Figure 4.14).

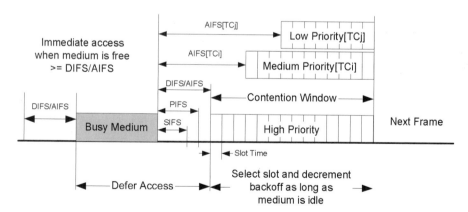

Figure 4.14 IEEE 802.11 EDCF

After the QoS stations wait for AIFS, each backoff sets a counter to a random number that is less than the size of the contention window (CW). Each traffic category has its own CW parameter. To achieve compatibility over legacy 802.11 WLANs, AIFS is set equal to DIFS. Similar to DCF, when the wireless medium is determined busy before the counter reaches zero, the backoff has to wait for the wireless medium to be idle for another AIFS period. If an unsuccessful transmission occurred, a new enlarged CW is computed by using the persistence factor (PF). For example, if PF=1, CW remains unchanged. If PF=2, CW applies a binary exponential backoff algorithm. Therefore, the key parameters (maintained in each traffic category) that enable priorities are AIFS, CW, and PF.

The transmitting stations can retain control of the wireless medium by using the short interframe spaces (SIFSs) and the virtual carrier sense mechanism (NAV). The station can transmit a number of packets that will look like a single instance of activity to other contending stations.

A QoS station has the option to implement up to eight transmission queues based on TCs that dictate the queue priority. In case there are two or more TCs in the station and some of their backoff counters reach zero at the same time, the station's scheduler grants the transmission opportunity to the queue with the highest TC priority.

EDCF can still result in collisions among stations that carry traffic with the same priority. A station is allowed a maximum transmission duration (transmission opportunity [TXOP]) as defined in the Beacon message.

The RTS and CTS mechanism defined in 802.11 DCF can also be implemented in the EDCF.

QoS in EDCF

EDCF combines a collision based channel access (CSMA/CA) and priority packet scheduling to priority CSMA/CA (see Figure 4.15). Using traffic category classification and priority CSMA/CA, EDCF delivers differentiated QoS services (prioritized QoS).

4.2.2.2   Hybrid Coordination Function (HCF)

HCF defines more rules than EDCF allowing more control of transmissions (see Figure 4.16). The HCF allocates bandwidth and transmission opportunities (TXOPs) using a hybrid coordinator (HC) that has the highest access priority. HCF uses a centralized polling based approach similar to PCF. During CP, each TXOP starts as defined by the EDCF rules—that is, after AIFS plus backoff time, or when the station receives from the HC a special poll frame, referred as the QoS Contention-Free (CF)-Poll. The HC sends this QoS CF-Poll after it waits for the wireless media to be idle for PIFS. The QoS CF-Poll specifies TXOP, which is the time interval during which the station has the right to transmit. During the CFP, only the polled station is allowed to transmit. The CFP ends after the time announced in the beacon frame or by a CF-End frame from the HC.

In order to provide quantitative QoS services, HCF requires a signaling process that informs the HC about the transmission requirements of each traffic stream at each station. Using this information, the HC will determine which stations need to be polled, when, and which TXOP should be granted. The detailed signaling process will be discussed in the next section. TXOP is granted per-station—that is, the HC does not specify which traffic stream should be transmitted on the channel. It is up to the station to select the traffic stream to be transmitted.

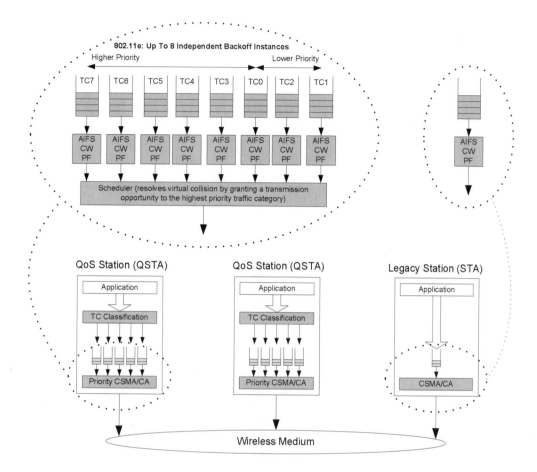

Figure 4.15 EDCF QoS Architecture

QoS in HCF

As shown in Figure 4.17, HCF contains the following QoS mechanisms: polling based channel access, QoS signaling, and traffic stream classification. These mechanisms enable the delivery of quantitative QoS services based on the applications' QoS requirements (TSPEC). However, IEEE 802.11e does not define the algorithms to set HCF parameters and it does not define the relationship between these parameters and the required QoS. Also, IEEE 802.11e does not present algorithms to compute the polling sequence and the time intervals provided in the HCF CF-Poll.

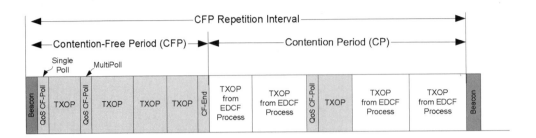

Figure 4.16 HCF CFP Repetition Interval

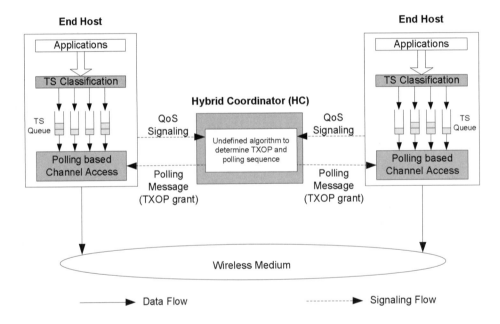

Figure 4.17 HCF QoS Architecture

4.2.3   QoS Signaling

The standard defines the QoS signaling which supports the operation of the QoS mechanisms. There are two forms of QoS signaling for traffic originating at the QoS stations: Queue State Indicator and Traffic Specification. QoS station transmits the message that contains Queue State or Traffic Specification to HC. QoS station can transmit the signaling message in three different ways.

1. *Transmit during CFP interval using HCF:* QoS station sends the signaling message during the CFP interval through the HCF polling process.
2. *Transmit during CP interval using EDCF:* The station sends the signaling message and contends with other stations for channel access.
3. *Transmit during CP interval using Controlled Contention (CC):* The station sends the signaling message during Controlled Contention Interval (CCI). The controlled contention process provides stations with the opportunity to request TXOPs by sending resource requests, without contending with other EDCF traffic. Each controlled contention instance starts when the HC sends a specific control frame that forces legacy IEEE 802.11 stations to set their NAV until the end of the controlled contention interval, thus they will refrain from transmission during the controlled contention interval (see Figure 4.18).

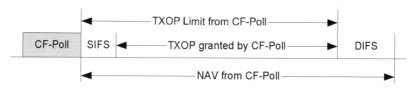

Figure 4.18 NAV Setting

4.2.3.1   Queue State Indicator Signaling

A QoS station transmits the queuing information of a traffic stream in the station to the HC. This will provide the information that is helpful to the HC to allocate the proper TXOP duration for the station to meet the traffic stream's demand. The algorithm that determines the TXOPs for each station using the queue information is not defined by the standard.

4.2.3.2   Traffic Specification Signaling

This signaling supports the resource reservation process in which a traffic stream in a QoS station transmits the signaling message that includes the traffic specification (i.e, bandwidth, delay requirement) to the HC. The signaling cooperates with the resource reservation process at higher layers (i.e., RSVP).

CHAPTER   5

HiperLAN

5.1 Introduction

The HiperLAN/2 set of standards was introduced by ETSI (the European Telecommunications Standards Institute) Project BRAN (Broadband Radio Access Networks). ETSI worked in collaboration with a HiperLAN2 Global Forum, which is a consortium of more than 50 companies established to market this new standard. In addition, HiperLAN/2 was developed in harmony with Japan's WLAN, referred to as HiSWANa. HiSWANa was developed by Japan's ARIB (Association of Radio Industries and Businesses) and its Multimedia Mobile Access Communications (MMAC) promotion association. Consequently, HiperLAN/2 and HiSWANa are very similar. Relatively few companies have products for HiperLAN/2 (e.g., Panasonic, and Amphion). HiperLAN/2 is designed to provide access to external IP, Ethernet, IEEE 1394, ATM, and 3G networks via an access point.

HiperLAN/2 provides data rates of up to 54 Mbps using Orthogonal Frequency Division Multiplexing (OFDM) radio technology for a range of up to 150 meters. This generation of standards supports both asynchronous data and applications that require QoS support. Notice that this standard, as other standards, defines signaling procedures required for the implementation of such QoS support. However, the algorithms that use this signaling information are not defined and are left for the developer.

The Media Access Control (MAC) protocol implements a form of dynamic time division duplex (TDD) and dynamic time division multiple access (TDMA) to provide connection oriented service and QoS support. The standard also supports station mobility with speeds of up to 10 ms.

The predecessor of HiperLAN/2, HiperLAN/1, was geared for ad hoc networking and asynchronous data. It applied the carrier-sense multiple access with collision avoidance (CSMA/CA) mechanism to resolve contention. HiperLAN/1 is considered a best effort system in spite of the fact that it included some preliminary mechanisms for QoS support.

5.2 Architecture

5.2.1   Network Topology

HiperLAN/2 distinguishes between a business environment and a residential environment. For a business environment the standard envisions a number of Access Points (APs) covering a certain area which may or may not overlap, depending on the business needs (Figure 5.1). This network topology is called the cellular access network configuration. An AP provider interconnects all mobile terminals (MTs) associated with it. All communication goes through the AP.

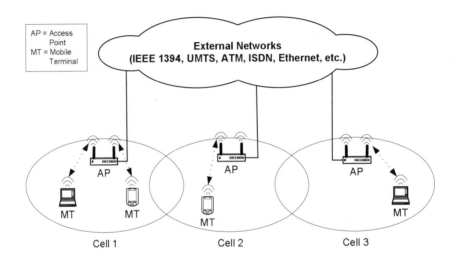

Figure 5.1   HiperLAN/2 Business Network (cellular access network configuration)

The residential environment is similar to the business environment. In addition, it can be operated in an ad hoc manner in which the MTs communicate with each other directly (see Figure 5.2).

The network topology for the residential environment is called an ad hoc LAN configuration. An ad hoc LAN configuration also requires a station called a central controller (CC), which provides network configuration control to all devices within the subnet. A CC in an ad hoc LAN configuration is similar to the AP in the cellular access network configuration. Unlike the dedicated hardware required for the AP, a CC is dynamically selected from the HiperLAN/2 mobile devices. When an MT which acts as a CC leaves the network, the CC responsibility can be handed over to another MT. HiperLAN/2 allows for multiple subnets and their respective CCs in the residence environment, similar to multiple cells and their respective APs in the business environment. Such multiple subnets can coexist since they operate at different frequencies.

HiperLAN/2 allows for two modes of operation: Centralized mode and Direct mode. In Centralized mode all traffic has to pass through the AP that manages the access to the wireless

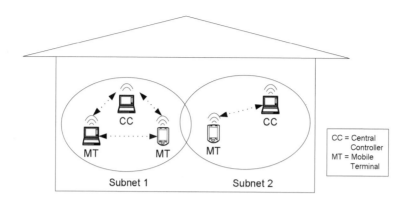

Figure 5.2   HiperLAN/2 Residential Network (ad hoc network configuration)

media. This traffic includes both traffic between mobile devices in the network and traffic between the mobile device and the outside network. In Direct mode the wireless media access is still managed by the CC. However, the traffic between the mobile devices is exchanged directly without going through the CC. Since the CC can also be connected to a core network, it is required that the CC will be able to work in both Direct and Centralized modes.

5.2.2   Protocol Stack

As shown in Figure 5.3, HiperLAN/2 defines the following three layers: the Convergence layer, the Data Link Control (DLC) with its MAC and other functionality, and the Physical layer.

The convergence layer provides the protocol interface between the upper layer and the DLC layer. The standard envisions that HiperLAN/2 provides wireless access to the external or core networks such as Internet Protocol (IP) networks, Asynchronous Transfer Mode (ATM) networks, 3G networks, and networks that use IEEE 1394 (Firewire) protocols. The data transport function of the convergence layer provides message format transformation (i.e., segmentation and reassembly function) between the higher layer and the DLC layer.

The data link layer contains three main entities: the Radio Link Control (RLC), the Error Control (EC), and Media Access Control (MAC). The user data plane function of the DLC receives the data packets from the upper layer (i.e., CL layer), provides the error control mechanism, and delivers the DLC data packets through the MAC. The RLC which is the control plane function of the DLC provides the Radio Resource Control (RRC), Association Control Function (ACF), and DLC Connection Control (DCC).

The HiperLAN/2 MAC of the AP/CC controls all transmissions over the wireless media. This includes uplink traffic from the station to the AP/CC and downlink traffic to the station. The MAC uses time division duplex (TDD) and dynamic time division multiple access (TDMA).

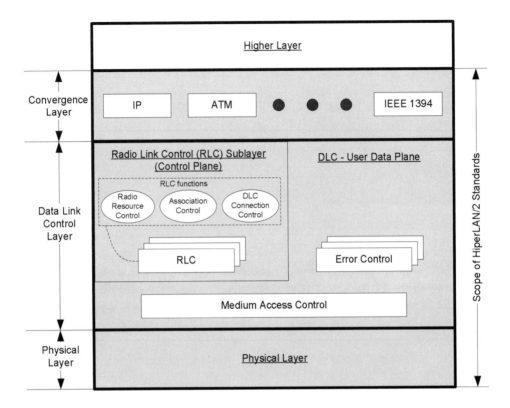

Figure 5.3   HiperLAN/2 Protocol Stack

5.3 Physical Layer

The physical layer supports several modes of transmissions: 6, 9, 12, 18, 27, 36, and 54 Mbps. Each mode employs a different radio modulation technique as shown in Table 5.1.

Table 5.1   HiperLAN/2 Radio Modulation Techniques

Modulation	Code Rate	Physical Layer Bit Rate (Mbps)
BPSK	1/2	6
BPSK	3/4	9
QPSK	1/2	12

Table 5.1 HiperLAN/2 Radio Modulation Techniques (Continued)

Modulation	Code Rate	Physical Layer Bit Rate (Mbps)
QPSK	3/4	18
16QAM	9/16	27
16QAM	3/4	36
64QAM (optional)	3/4	54

HiperLAN/2 operates at 5 GHz frequency spectrum as shown in Figure 5.4. In the U.S., 300 MHz bandwidth in U-NII band has been allocated, in Japan 100 MHz (with sharing rule) bandwidth has been allocated, and in Europe 455 MHz bandwidth has been allocated in license exempt band. The channel spacing is 20 MHz, which allows high bit rates per channel but still has a reasonable number of channels in the allocated spectrum (e.g., 19 channels in Europe).

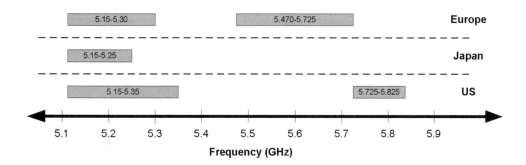

Figure 5.4 HiperLAN/2 Frequency Spectrum Allocation

HiperLAN/2 uses orthogonal frequency division multiplexing (OFDM). The transmission unit is a burst, which consists of a preamble part and a data part, where the latter could originate from each of the transport channels within the DLC.

HiperLAN/2 defines Dynamic Frequency Selection (DFS), which allows sharing of the wireless media among several HiperLAN/2 networks. The AP chooses frequencies based on a frequency selection process considering interference measurements. In addition, the AP/CC will adapt its transmission mode (speed, code, and modulation) based on its measurements of the transmission link quality. Moreover, the mobile stations, based on their measurements, may suggest the preferred transmission mode.

5.4 Data Link Control (DLC) Layer

As mentioned earlier, the DLC layer contains three sublayers: Error Control (EC), Radio Link Control (RLC), and Media Access Control (MAC). In addition, there are logical and transport channels as shown in Figure 5.5. Logical channels are identified by the type of message they carry while transport channels are identified by the message format and the channel access method. The control messages and user data messages (which originate from the DLC or higher layer) are mapped to the appropriate logical channel based on their contents. Then the messages are passed to the transport channel in order to construct the MAC frame and receive the appropriate channel access. In the next subsections we introduce the functionality of each DLC sublayer, the MAC frame, and the transport and logical channels. Then we describe MAC protocol examples.

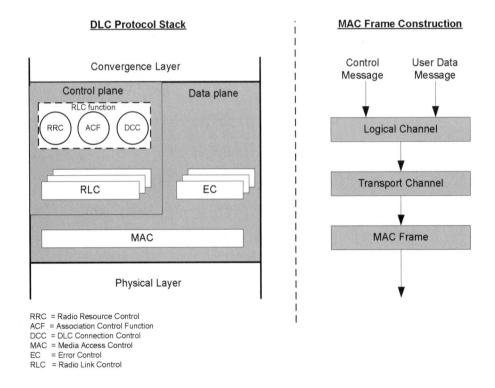

Figure 5.5 DLC Protocol Stack and MAC Frame Construction

5.4.1   Radio Link Control (RLC)

The RLC manages the network and thus exchanges control data between the AP and the MTs. The AP can have multiple RLC instances, where each RLC instance associates with an MT. There is only one RLC instance in an MT.

The RLC has three functions: Radio Resource Control (RRC), Association Control Function (ACF), and DLC Connection Control (DCC).

The RRC is responsible for detecting and efficiently using the available radio resources. It manages handover, dynamic frequency selection, station alive/absent, power saving, and power control. The RRC selects the frequency range on which the communication will be conducted and, when needed, it decides to move to a different frequency range. This is done based on the AP/CC own channel measurements as well as the mobile station channel measurements. To preserve the battery power of the stations, the RRC defines when and for how long to put the MTs into sleep mode. The lengths of sleep intervals are negotiated between the AP/CC and the MTs.

The ACF manages the association process for authentication, key management, association, disassociation, and encryption. A station that wants to communicate over the wireless media is required to be associated with an AP/CC. This is due to the fact that the AP/CC allocates wireless resources for each associated station. Moreover, the MAC is centrally controlled by the AP/CC in both Centralized Mode and Direct Mode. When a new MT joins the network by performing the association process with the AP's ACF, the ACF assigns a unique MAC Identification number (MAC ID) to the MT. The ACF also maintains information of the MT's capabilities in terms of whether encryption and authentication are performed, and which encryption and authentication methods are employed.

The DCC sets up, maintains, and releases user connections. If any kind of QoS support is requested, the connection has to supply the QoS parameters. The specification of these parameters is not part of the HiperLAN/2 standard. The connection setup includes procedures for centralized mode or direct mode, procedures for connection release, and procedures for joining and leaving multicast groups.

5.4.2   Error Control (EC)

The EC is responsible for detection and recovery from transmission errors on the wireless media. It also ensures in-sequence delivery of packets. Each connection is supported by an instance of the EC.

The EC is based on an Automatic Repeat Request (ARQ) algorithm. Additional error correction techniques can be employed. The ARQ scheme is based on a selective repeat mechanism in which the receiver requests that the sender retransmit packets determined to be erroneous. In addition, the EC module provides several modes to improve the transmission reliability:

1. *Acknowledged mode:* The EC will retransmit the acknowledgments from the receiver. Low latency can be maintained by a mechanism that discards packets based on time information.

2. *Repetition mode:* The EC will repeat the transmission. No acknowledgments are available in this mode. Thus, this mode is typically used for broadcast data.

3. *Unacknowledged mode:* This is for unreliable, low latency transmission without retransmissions.

Therefore, unicast data can be sent using either the Acknowledged mode or the Unacknowledged mode. Broadcast data can be sent using either the Repetition mode or the Unacknowledged mode. Multicast data can be sent in the Unacknowledged mode or can be piggybacked into other unicast transmissions.

5.4.3   Media Access Control (MAC)

The HiperLAN/2 MAC that resides at the AP/CC controls all the transmissions over the wireless media. This includes 1) uplink transmissions from the MTs to the AP/CC, 2) downlink transmission from the AP/CC to the MTs, and 3) direct transmission among the MTs. Direct transmission is mandatory for the residential or ad hoc configuration. Therefore, the MAC deploys time division duplex (TDD) and dynamic time division multiple access (TDMA). The channel is structured into MAC frames as shown in Figure 5.6. The MAC frame has a fixed duration of 2 ms. Each MAC frame starts with the Broadcast Channel (BCH) duration.

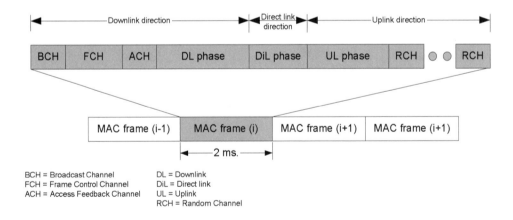

Figure 5.6 MAC Frame Structure

As shown in Figure 5.7, the frame structure slightly differs if the AP/CC has a sectored antenna. In this case, each phase is repeated in time, one for each sector. The use of Direct Link (DiL) with sectored antennas is not specified.

The MAC frame consists of: Broadcast Channel (BCH) duration, Frame Control Channel (FCH) duration, Access Feedback Channel (ACH) duration and at least one Random Channel (RCH) duration. If there is transmission between the AP/CC and the MTs, the Downlink (DL)

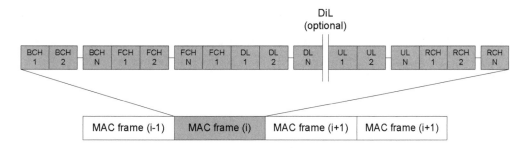

Figure 5.7   MAC Frame Structure for Sectored Antennas

phase and/or Uplink (UL) phase are included in the MAC frame. If there is transmission among the MTs (direct mode), the Direct Link (DiL) phase is also included. The BCH duration is fixed. The duration of the FCH, DL, DiL, and UL phases and the number of RCHs are dynamically determined by the AP/CC.

The BCH, FCH, ACH, and RCH contain control messages, whereas the DL phase, DiL phase, and UL phase mostly contain user data and certain control messages. The standard defines the message format through the transport and logical channels. There are six transport channels (with three letter abbreviations):

1. Broadcast Channel (BCH)
2. Frame Channel (FCH)
3. Access Feedback Channel (ACH)
4. Long Transport Channel (LCH)
5. Short Transport Channel (SCH)
6. Random Access Channel (RCH)

There are ten logical channels (with four letter abbreviations):

1. Broadcast Control Channel (BCCH)
2. Frame Control Channel (FCCH)
3. Random Access Feedback Channel (RFCH)
4. User Data Channel (UDCH)
5. User Multicast Channel (UMCH)
6. User Broadcast Channel (UBCH)
7. RLC Broadcast Channel (RBCH)
8. Dedicated Control Channel (DCCH)
9. Link Control Channel (LCCH)
10. Association Control Channel (ASCH)

Figure 5.8 illustrates the MAC frame structure, the transport and logical channels and their relationships. All messages are mapped to specific logical channels based on their content and their functionality. The messages are constructed according to a format defined by the logical channel. Then the messages will pass to the transport channel to receive the appropriate transport services.

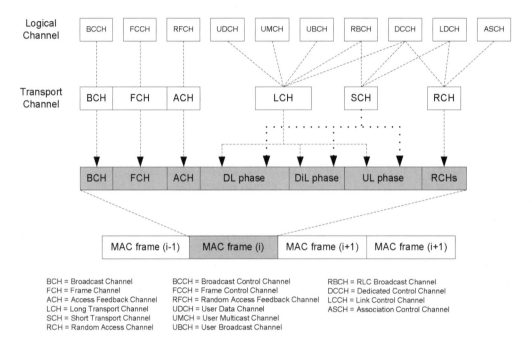

BCH = Broadcast Channel   BCCH = Broadcast Control Channel   RBCH = RLC Broadcast Channel
FCH = Frame Channel   FCCH = Frame Control Channel   DCCH = Dedicated Control Channel
ACH = Access Feedback Channel   RFCH = Random Access Feedback Channel   LCCH = Link Control Channel
LCH = Long Transport Channel   UDCH = User Data Channel   ASCH = Association Control Channel
SCH = Short Transport Channel   UMCH = User Multicast Channel
RCH = Random Access Channel   UBCH = User Broadcast Channel

Figure 5.8   MAC Frame Structure, Transport Channel, and Logical Channel

Each MT's MAC is uniquely identified by a MAC ID which was assigned by the AP's RLC during the association process. The HiperLAN/2 MAC is connection oriented. Before an application can transmit, it requires the MT to establish a connection with the AP's RLC. This RLC assigns to this connection a DLC Connection ID (DLCC ID). In the centralized mode (CM), the DLC User Connection ID (DUC ID), which is obtained as a combination between the MAC ID and the DLCC ID, uniquely identifies the connection in the cell. In distribution mode (DM), the definition of DUC ID is slightly different. DUC ID, which identifies the connection in DM mode, is the combination of source MAC ID, destination MAC ID, and DLCC ID. DUC ID is also used for identifying the control channels. DUC ID is essential for the AP to allocate bandwidth to a specific connection (both user data and control).

5.4.4   Logical Channel

A logical channel is viewed as a generic term for distinct data paths. Logical channels are defined by the type of information they carry.

5.4.4.1   Broadcast Control Channel (BCCH)

The BCCH is used in the downlink direction and carries BCCH information about the entire wireless cell to all the stations within the cell. The BCCH message is only forwarded to BCH transport channel and eventually is transmitted on the BCH. BCCH contains: Net ID (the network or the cell identifier), AP ID (AP identifier), AP TX level (indicating the transmission power of the AP), AP RX UL Level (indicating the expected reception power level of the AP). Moreover, the BCCH also contains a pointer or a location and the length of the FCH duration and RCH duration in the MAC frame as shown in Figure 5.9. The support provided by BCCH is mandatory for APs and CCs. The mobile stations should be able to interpret the BCCH.

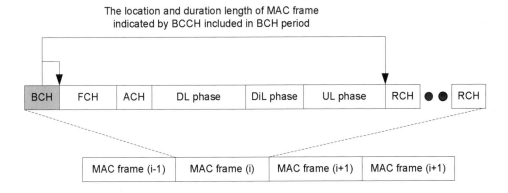

Figure 5.9   The Pointers in BCCH

5.4.4.2   Frame Control Channel (FCCH)

The FCCH is used in the downlink direction and carries information that describes the structure of the MAC frame visible at the air interface (see Figure 5.10). FCCH is also called Resource Grants (RGs). An RG, which is per connection basis (DUC ID), corresponds to a number of transport channels (the number of LCHs or SCHs), the PHY modes to be used, and the location in the frame where the reception/transmission will take place. The connection can be a control connection (RBCH, DCCH, LCCH) or a user data connection (UDCH, UBCH, UMCH). In addition, FCCH includes announcements of the empty parts in the MAC frame. FCCH becomes an Information Element (IE) of the FCH transport channel. The FCH can contain multiple IEs. Therefore, the FCH's duration is variable. The support of the FCCH is mandatory for APs and CCs. MTs are required to be able to interpret the FCCH.

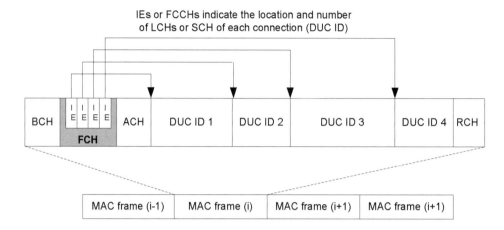

IEs or FCCHs indicate the location and number
of LCHs or SCH of each connection (DUC ID)

| BCH | FCH (E E E E) | ACH | DUC ID 1 | DUC ID 2 | DUC ID 3 | DUC ID 4 | RCH |

| MAC frame (i-1) | MAC frame (i) | MAC frame (i+1) | MAC frame (i+1) |

Figure 5.10 Frame Control Channel (FCCH) in FCH

5.4.4.3 Random Access Feedback Channel (RFCH)

The RFCH carries information for stations that have used the RCH in the previous MAC frame. This information contains the result of their access attempts. The RFCH message, which is transmitted once per MAC frame, is included in the ACH (see Figure 5.11). The support of the RFCH is mandatory for APs and CCs. MTs are required to be able to interpret the RFCH.

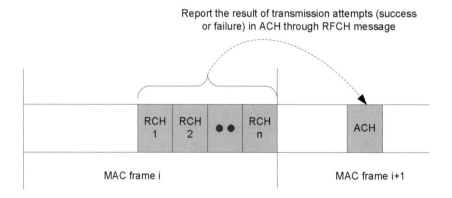

Report the result of transmission attempts (success
or failure) in ACH through RFCH message

| RCH 1 | RCH 2 | ● ● | RCH n | | ACH |

| MAC frame i | MAC frame i+1 |

Figure 5.11 ACH Operation

5.4.4.4 RLC Broadcast Channel (RBCH)

The RBCH is used in the downlink direction in Centralized Mode (CM) and may be used by an MT when it communicates without the AP/CC in Direct Mode (DM). RBCH carries broadcast control information about the whole radio cell. RBCH transmission is triggered by the AP/CC in the centralized mode and is determined by the originating MT in the direct mode. The following information may be included in the downlink RBCH: broadcast RLC messages, transmission of the assigned MAC ID to a non-associated mobile terminal, convergence layer ID information, and encryption related information. Figure 5.12 shows the RBCH frame construction process. The support of the downlink RBCH in AP/CC and MTs is mandatory.

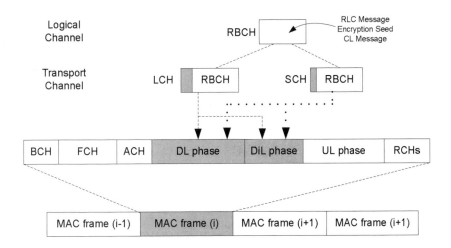

Figure 5.12 RBCH Frame Construction Process

5.4.4.5   Dedicated Control Channel (DCCH)
The DCCH is used to carry RLC messages. It is used in the uplink, downlink, and Direct Mode communication. The DCCH may also be used for the transmission of Resource Requests (RRs) in the uplink direction. DCCH is transmitted through LCH or SCH or RCH transport channel. Figure 5.13 shows the DCCH frame construction process. The support of the DCCH in both MTs and AP/CC is mandatory. It is required in Direct Mode if this feature is supported.

5.4.4.6   User Broadcast Channel (UBCH)
The UBCH is used to transmit user broadcast data. The UBCH can be sent by the AP/CC in the downlink and by an MT in the direct link. The UBCH is transmitted in repetition or unacknowledged mode. UBCH is transmitted through LCH transport channel. UBCH frame construction process is shown in Figure 5.14. The support of the UBCH in both MTs and AP/CC is mandatory. It is required in Direct Mode if this feature is supported.

5.4.4.7   User Multicast Channel (UMCH)
The UMCH is used to transmit user multicast data. The UMCH can be sent by the AP/CC in the downlink and by an MT in direct link. The UMCH is transmitted in repetition or unacknowledged mode. UMCH is transmitted through LCH transport channel. UMCH frame construction process is shown in Figure 5.14. The support of the UMCH in both MTs and AP/CC is mandatory. It is required in Direct Mode if this feature is supported.

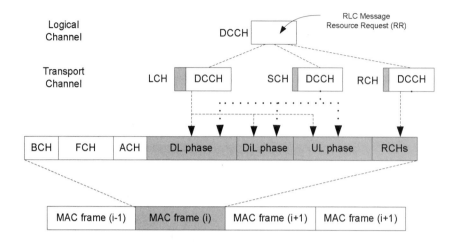

Figure 5.13 DCCH Frame Construction Process

5.4.4.8   User Data Channel (UDCH)

The UDCH is used to transmit user data between the AP and the MTs in Centralized Mode (CM) or between two MTs in Direct Mode (DM). A UDCH is always granted together with zero or more transport channels for a connection that is announced in the RG in the FCCH. UDCH is transmitted through LCH transport channel. UDCH frame construction process is shown in Figure 5.14. The support of UDCH for uplink and downlink is mandatory for both MTs and AP/CC. The support of the UDCH for Direct Mode is mandatory if this feature is supported.

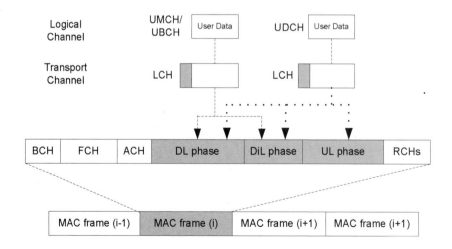

Figure 5.14 UMCH/UBCH/UDCH Frames Construction Process

5.4.4.9   Link Control Channel (LCCH)

The LCCH is used in both the uplink and downlink to transmit error control information such as ARQ feedback. It is used in either the Centralized or Direct Mode. The LCCH is also used in the uplink transmission of the RRs. LCCH is transmitted through the SCH or RCH transport channels. LCCH frame construction process is shown in Figure 5.15. The support of the LCCH for uplink and downlink is mandatory for both MTs and AP/CC. The support of the LCCH for Direct Mode is mandatory if this feature is supported.

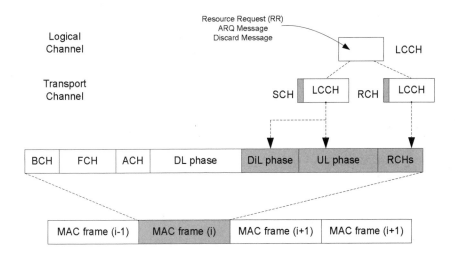

Figure 5.15 LCCH Frame Construction Process

5.4.4.10   Association Control Channel (ASCH)

The ASCH is only used in the uplink and carries new association and handover request messages. These messages are sent only by the MTs that are not associated to an AP/CC. ASCH is transmitted through the RCH transport channel. ASCH frame construction process is shown in Figure 5.16. The support of the ASCH is mandatory for AP/CC and MTs.

5.4.5   Transport Channels

The various logical channels are mapped into corresponding transport channels that describe the basic message format.

1. *Broadcast Channel (BCH):* The BCH carries the BCCH in the downlink direction. Its support is mandatory for AP/CC and stations.
2. *Frame Channel (FCH):* The FCH carries the FCCH. It is broadcasted in the downlink direction and is mandatory for all AP/CC and stations.
3. *Access Feedback Channel (ACH):* The ACH carries RFCH in the downlink. Its support is mandatory for AP/CC and stations.

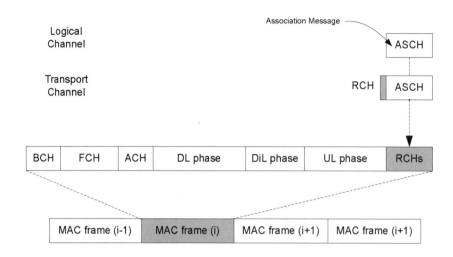

Figure 5.16 ASCH Frame Construction Process

4. *Long Transport Channel (LCH):* The LCH carries user data for the connections related to the granted UDCHs, UBCHs, and UMCHs, as well as control information for the connections related to the DCCH and RBCH. Its support is mandatory for AP/CC and stations.

5. *Short Transport Channel (SCH):* The SCH carries short control information for the DCCH, LCCH, and RBCH. Its support is mandatory for AP/CC and stations.

6. *Random Access Channel (RCH):* The RCH is defined for the purpose of giving a station the opportunity to send control information to the AP/CC when it has no granted SCH available. It can carry RRs, ASCH, and DCCH data. Its support is mandatory for AP/CC and stations.

5.4.6   Resource Request and Resource Grant

The standard defines the resource request (RR) and resource grant (RG) signaling mechanism to support the bandwidth allocation algorithm, which plays an important role in QoS support. The standard does not define the bandwidth allocation algorithm. Figure 5.17 shows a simplified diagram of the bandwidth allocation input and output.

Because HiperLAN/2 MAC is connection oriented, bandwidth allocation is performed per connection basis. Each RR and RG serves for an individual connection. In order for the AP/CC to allocate resources effectively it needs to know the bandwidth demand of each connection represented by the queue status (i.e., queue length). Thus, the stations will report their queue status in the RR messages to the AP/CC. Using the queue information received, the QoS parameters, and the available network resources, the bandwidth allocation algorithm which resides at the

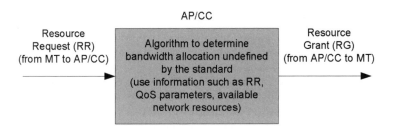

Figure 5.17 Bandwidth Allocation

AP/CC will determine the RG. The RG indicates when a connection is allowed to transmit and how many packets can be sent. The RG allocation algorithm which is not defined by the standard needs to be determined by the system developer.

The RR can operate in several ways:

1. *Polling Resource Request* (Figure 5.18): An MT initiates the RR by listening to the poll. First, the AP/CC indicates the polling for an RR by setting the RR Poll bit in the FCCH message. Then, the MT will transmit the RR using LCCH with SCH transport channel on the uplink.

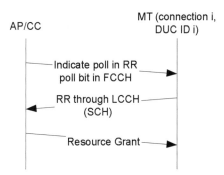

Figure 5.18 Polling Resource Request

2. *Unpolling Resource Request* (Figure 5.19): An MT initiates the RR, which is sent through DCCH with SCH transport channel (collision-free transport) and/or LCCH with RCH transport channel (collision-based transport). AP/CC can control the access delay of the RCH by changing the number of contention slots (i.e., the number of RCHs). If a collision occurs, the stations are informed in the ACH of the next MAC frame. Then, the station backs off a random number of access slots.

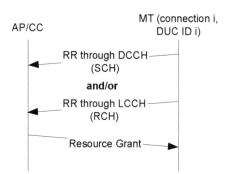

Figure 5.19 Unpolling Resource Request

5.5 Convergence Layer

The convergence layer (CL) maps service requests from the higher layers to the service offered by the DLC. It converts higher layer packets of fixed (ATM) or variable (Ethernet) length into fixed-length packets, referred to as Service Data Unit (SDU), that are used within HiperLAN/2. For example, CL will map the 802.1p priority field into a HiperLAN/2 priority class.

5.6 QoS support

HiperLAN/2 defines several QoS mechanisms to provide QoS support for multimedia applications. HiperLAN/2 manages the resources based on a connection-oriented approach which enables the finest granularity of QoS support (per flow and quantitative QoS services). In this section, we describe each QoS mechanism provided by HiperLAN/2.

5.6.1 Classification

Each connection is tagged with DUC ID. In centralized mode (CM), DUC ID consists of MAC ID and DLC connection ID (DLCC ID). In distributed mode (DM), DUC ID consists of source MAC ID, destination MAC ID and DLCC ID. DUC ID enables classification. Classification identifies packets based on DUC ID and forwards them to the appropriate queue. Therefore, HiperLAN is capable of per-flow classification which supports per-flow QoS services.

5.6.2 Channel Access

As described in Section 5.4.3, the channel access is TDD/TDMA, which is controlled by the AP/CC through FCCH. Since TDD/TDMA is a collision-free channel access scheme which provides tight channel access control, HiperLAN/2 can provide QoS support for applications with strict QoS requirements.

5.6.3   Packet Scheduling

The packet scheduling algorithm 1) allocates bandwidth for connections in terms of the number of packets (i.e., number of LCHs) and 2) determines when a connection is allowed to transmit. This information is defined in the FCH IEs (FCCH). The packet scheduling algorithm uses the RR and RG, which are described in Section 5.4.6. The standard does not define the packet scheduling algorithm. It just defines the signalling mechanism such as RR and RG.

Figure 5.20 summarizes the HiperLAN/2 QoS architecture which provides the necessary mechanisms to deliver per-flow quantitative QoS services.

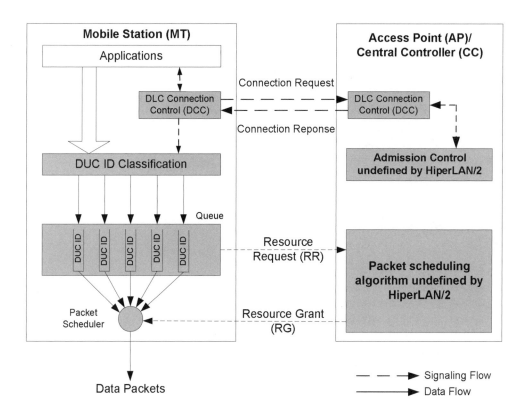

Figure 5.20 HiperLAN/2 QoS Architecture

HomeRF

6.1 Introduction

HomeRF 2.0 standard was developed by the HomeRF Working Group (HRFWG) formed by companies such as Siemens, Proxim, National Semiconductor, Xilinx, Motorola, and AT&T. The group goal is to establish the mass deployment of interoperable wireless networking access devices to both local content and the Internet for voice, data, and streaming media in consumer environments. Relative to IEEE 802.11, few products have been announced.

Companies such as Cayman Systems, Compaq, Intel, Motorola, and Proxim have announced HomeRF 2.0 based communication cards. Utilizing the HomeRF QoS support for voice, companies such as Motorola and Siemens introduced products that include voice delivery. Motorola introduced the Simplefi system, which allows users to transmit Internet digital audio to the existing stereo equipment. Siemens introduced Voice Data Gateway and HomeRF based phones.

HomeRF 2.0 specification defines a common interface that supports wireless voice and data networking in the home. The specification allows 800 kbps rate for isochronous voice and rates of 1.6, 5, and 10 Mbps for asynchronous data transfer. HomeRF specification is backwards compatible with HomeRF 1.2, which operates at rates of 0.8 Mbps and 1.6 Mbps. HomeRF targets to support a wide range of applications by defining the following three types of data services:

1. *Asynchronous data service:* It provides best effort service for asynchronous data applications such as file transfer and email.
2. *Priority asynchronous data service:* It provides priority service for asynchronous data applications with loose QoS requirements, such as streaming media (video, audio).
3. *Isochronous data service:* It provides support for applications with strict QoS requirements (i.e., delay, delay jitter) such as interactive voice and cordless phone.

HomeRF 1.x standard did not describe roaming procedures in terms of device hand-offs and resolution of multiple address conflicts. HomeRF 2.0 added the possibility of an extended network that is an aggregation of individual wireless HomeRF networks associated with the same network ID number. Such a network is managed by roaming capable Connection Points (CPs). The standard also defines roaming procedures which include hand-offs, addresses, and data transfers between roaming devices.

6.2 Architecture

6.2.1 Network Topology

Before we describe the HomeRF network topology, we would like to introduce the HomeRF devices and their functionalities. The HomeRF standard categorizes the HomeRF devices into three main types, based on the type of application the devices support:

1. *Asynchronous data device* (also called A-node): These devices, such as PC, Laptop, and PDA, support asynchronous data services (i.e., file transfer, email).
2. *Streaming data device* (also called S-node): These devices, such as audio headsets, support priority asynchronous data services (i.e., audio or video streaming).
3. *Isochronous data device* (also called I-node): These devices, such as cordless phones, support isochronous data services (i.e., telephone call).

Some devices can support multiple services. For example, AI-nodes combine the functionalities of A-node and I-node. SI-nodes combine the functionalities of S-node and I-node. SA-nodes, such as Internet appliances, combine the functionalities of S-node and A-node. The CP provides service management for the A-node, S-node and I-node. The CP has the following functionalities:

- *A-node management:* It supports A-nodes with power saving and provides data access between the A-node and the Internet or PCs on other network segments.
- *S-node management:* It provides session setup for the S-node and assigns priority channel access to the S-node.
- *I-node management:* It provides connection setup for the I-node, allocates the I-node dedicated network resources (i.e., bandwidth), and provides connectivity between the I-node and the Public Switch Telephone Network (PSTN).

The HomeRF standard defines two main network topologies: Ad hoc Network and Managed Network. An ad-hoc network (see Figure 6.1) is a distributed network (similar to the ad hoc network concept in IEEE 802.11) in which HomeRF devices can establish peer-to-peer communication. An ad hoc network includes only A-nodes. There is no CP in an ad hoc network. A-nodes in an ad hoc network cannot operate in the power saving mode.

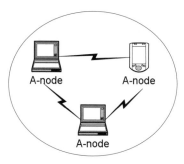

Figure 6.1   Ad Hoc Network

The Managed Network (see Figure 6.2) is a network managed by the CP. The HomeRF devices can either establish peer-to-peer communication with each other or establish communication through the CP depending on the type of the HomeRF devices, for example:

- *A-node:* Peer-to-peer communication between the A-nodes, or communication to the Internet or PCs on other network segments through the CP.

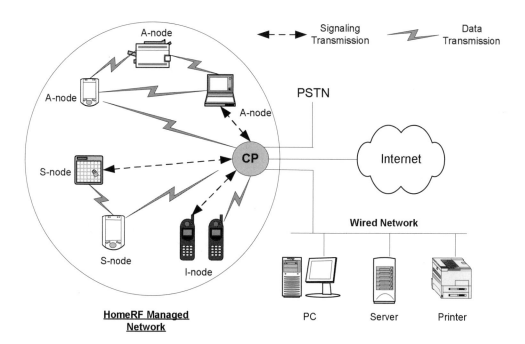

Figure 6.2   Managed Network

- *S-node:* Peer-to-peer data communication between the S-nodes, or communication with the CP for signaling purposes (i.e., session setup, priority assignment) or for data transfer to the Internet (i.e., streaming video from the Internet).
- *I-node:* Communication to PSTN or PC through the CP for transmission of both data and signaling. Peer-to-peer communication between I-nodes is not allowed.

6.2.2   Protocol Stack

As mentioned earlier, HomeRF protocols use best effort asynchronous data service, priority asynchronous data service, and isochronous data service to support simultaneous sessions for audio, video, and data. Both the asynchronous and priority asynchronous data services are packet switched using a carrier sense multiple access (CSMA/CA) media access control (MAC). The isochronous data service is for voice that requires strict jitter and latency and is supported by a circuit switched, time division multiple access (TDMA) MAC. HomeRF supports TCP/IP and User Datagram Protocol (UDP/IP) traffic as well as Digital Enhanced Cordless Telecommunications (DECT) traffic (see Figure 6.3). DECT, which is a cordless phone standard mainly used in Europe, uses a TDMA based MAC (i.e., assigns users with predetermined time slots). DECT provides speeds of up 2 Mbps. The implementation of the HomeRF MAC differs from the DECT MAC. However, because the services it provides are similar, the DECT higher layers can be hosted on the HomeRF MAC with minimal modification. The use of the DECT protocol enables HomeRF to support call setup for the isochronous data service and to interoperate with the PSTN.

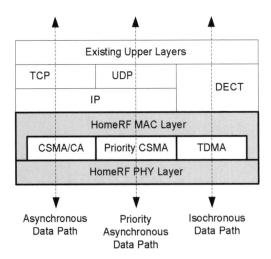

Figure 6.3   HomeRF Protocol Stack

6.3 Physical Layer

The PHY layer is based on Frequency Hopping Spread Spectrum (FHSS) and operates in the 2.4 GHz frequency band. It provides 75 1 MHz channels for 1.6 Mbps data and voice communication, and 15 5 MHz channels for 5 and 10 Mbps data communication.

HomeRF specifies an adaptation mechanism to ensure that two adjacent hops will not both be within the frequency range where interference may occur. When interference is detected, the hopping series set is examined to find out if two consecutive hops are both within the interference range. If such hops are found, then an attempt is made to switch these hops with hops that are outside the interference range. For example, if a microwave oven interferes, the hopping series will be adjusted to avoid the microwave frequency.

6.4 Media Access Control (MAC)

The standard organizes the communication channel into a superframe structure that includes two periods—contention period and contention-free period. The duration of the superframe is fixed and has two possible values: 20 ms when the superframe contains only a contention period (see Figure 6.4A) and 10 ms when the superframe contains both contention and contention-free periods (see Figure 6.4B). Each superframe begins with a hopping period (HOP) where HomeRF devices change their hopping frequency. When there is a contention-free period, the superframe contains two subframes—contention-free period for voice transmissions and optional contention-free period for voice retransmissions (see Figure 6.4B).

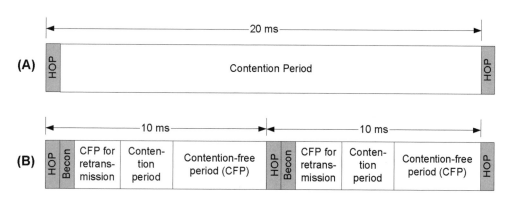

Figure 6.4   HomeRF Superframe Structure

6.4.1   Channel Access

The standard defines two channel access schemes for each period (contention and contention-free):

> **1.** *Contention period:* The channel access is CSMA/CA with added priority access and time reservation mechanisms. The priority access feature provides priority access to the streaming media transmission over the asynchronous data transmission. Therefore, this channel access is used by A-nodes and S-nodes.

2. *Contention-free period:* The channel access is TDMA with the re-transmission option. Since this channel access provides isochronous data transmission service (i.e., voice), it is used by the I-nodes (e.g., cordless phones).

As shown in Figure 6.5, the contention period includes asynchronous data transmission where the streaming media sessions receive priority. The contention-free period includes voice transmissions and optional voice retransmissions.

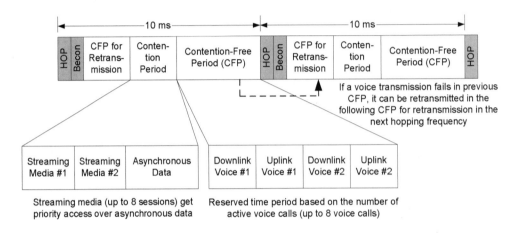

Figure 6.5   Channel Access

6.4.2   Contention-based CSMA/CA with Priority Access and Time Reservation

Contention-based CSMA/CA with priority access and time reservation supports two modes of asynchronous data access: legacy non-prioritized data access and priority asynchronous data access. In case of non-prioritized access, stations that have data to transmit access the wireless medium in an uncoordinated way. Consequently, this access has the lowest access priority. In the priority access, the stations access the wireless channel based on a priority access value determined by the CP. S-nodes will be granted priority access whereas A-nodes will access through the non-prioritized service. The CSMA/CA mechanism supports a multi-rate data service of 800 kbps, 1.6 Mbps, 5 Mbps, and 10 Mbps.

Similar to IEEE 802.11 (see Chapter 4), stations avoid collisions by listening to transmissions on the wireless medium. The station will transmit only if it senses that the medium is free for a certain period of time referred to as Interframe Space (IFS). Then, the station randomly selects a time slot within the contention window to start transmission. A collision is discovered by the fact that the sending station does not receive the acknowledgment from the receiving station. In this case, the station will double the contention window size and retransmit data by reselecting the time slot from the new contention window. The IFS is similar to IEEE 802.11:

- SIFS (short IFS) is the shortest time between two consecutive packet transmissions and used when there is no need to perform a Clear Channel Assessment (CCA) procedure before initiating a transmission.
- DIFS is the shortest time between the last bit of a sequence of a complete data transmissions and the first bit of a new packet sequence. This interval includes the time to recover from any prior transmission, perform a CCA procedure, and start a transmission. DIFS definition in HomeRF standard is slightly different from DIFS definition in IEEE 802.11. DIFS in HomeRF includes the randomized time slot duration of the contention window, which is not included in IEEE 802.11.

In addition, this MAC supports time reservation which enables the uninterrupted transmission of a sequence of multiple data packets. Stations that listen to the exchange of time reservation transmissions will not attempt transmissions. Hence, providing a contention-free medium is for the stations participating in this time reservation exchange. Time Reservations are used to signal that the time reservation is made, canceled, or requested. An optional Request-To-Send (RTS) and Clear-To-Send (CTS) packet exchange (similar to IEEE 802.11) can be used for the reservation process in case of hidden nodes. In the hidden node case, the RTS-CTS exchange ensures that packets at the source vicinity as well as the destination vicinity will defer for transmission during the reserved period of time.

The standard also defines CSMA/CA with priority access (see Figure 6.6), which can support up to eight priorities. Priority access can be achieved through reserved time slots (slot #1 up to slot #8) in the contention window. The CP assigns these slots. Priority access is used by streaming media applications. A streaming media application with stricter QoS requirements will be assigned a lower slot number which translates in a shorter DIFS. Other asynchronous data sessions select the slot number of the contention window excluding the reserved slots. Therefore, the streaming media with the lowest assigned slot number will have the highest priority access. Because the channel access scheme uses the CSMA/CA mechanism, the time delay is variable and cannot be guaranteed.

6.4.3   Reserved Time TDMA with Retransmission Option

Reserved time TDMA with re-transmission option operates in the contention-free period and is available only in a CP managed wireless network. The CP manages the I-nodes' access to the wireless medium by sending special control packets, or beacons, that contain TDMA information. Because all I-nodes' data transmission gets through the CP, there are two directions of transmission: uplink transmission (from I-node to CP) and downlink transmission (from CP to I-node). This MAC mode is connection oriented—that is, each voice call originated at an I-node is required to establish a connection with the CP. The CP will allocate a downlink-uplink pair of time slots to each call.

Figure 6.7 shows a potential scenario of HomeRF operation. Figure 6.7A shows the channel allocation when only data are communicated using the contention based protocol. The cycle is 20 ms long and is used by several data packets and their respective acknowledgments (ACKs)

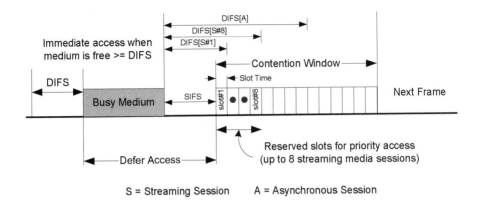

S = Streaming Session   A = Asynchronous Session

Figure 6.6   CSMA/CA with Priority Access

(Figure 6.7A). In Figure 6.7B, two voice conversations are added and the cycle is reduced to 10 ms. The Beacon, which is a management control packet, starts each cycle and allocates downlink and uplink TDMA time slots. After each cycle the Frequency Hopping radio hops to the next frequency of the frequency series.

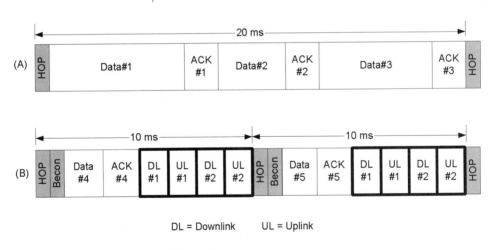

DL = Downlink   UL = Uplink

Figure 6.7   Example of HomeRF Operation

The connection-oriented TDMA protocol is referred to as the "single retransmission after hop" retry protocol (Figure 6.8). This protocol supports high quality voice transmissions (e.g., DECT based devices) that request bounded delays. The CP allocates time slots to every active connection. This allocation may change if other connections are terminated. The TDMA acknowledgment is a simple piggyback acknowledgment inserted in a later TDMA packet in the reverse direction. If the CP does not receive a valid TDMA packet or if the TDMA packet does not include an acknowledgment, then the CP schedules a retry time slot for the connection. This protocol provides bounded time delay because each time sensitive connection is provided dedicated bandwidth in both directions (from/to the CP) (i.e., dedicated TDMA slots in every cycle) and dedicated bandwidth in the succeeding frame if a retransmission is required. A retransmission, if needed, is done in the consecutive cycle which is less than 10 ms later.

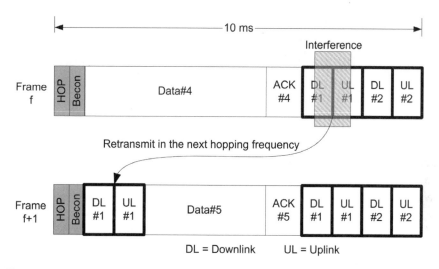

Figure 6.8   TDMA with Retransmission

This protocol also provides support for a connectionless broadcast service used by broadcasting applications whose packets do not require acknowledgments. Such applications include I-node pages, cadence ringing, caller ID, and voice announcement. This service is provided by only using the downlink time slots.

6.5 QoS Support

HomeRF standard explicitly defines the QoS service it can support. As mentioned at the beginning of this chapter, HomeRF offers three main services: asynchronous data service, priority asynchronous data service, and isochronous data service. In this section, we elaborate the QoS aspects of each service. We will describe the service characteristics and the QoS mechanisms involved.

6.5.1   Isochronous Data Service

This service is designed for applications (i.e., PSTN call, Intercom call, PC call, Conferencing) that originate at an I-node which require strict QoS control in terms of bandwidth, delay, and delay jitter. To achieve this service, the following service characteristics and QoS mechanisms are involved:

1. *Connection-oriented service:* An I-node sends the connection request to the CP. The CP assigns the connection ID and allocates the slot position in the contention-free period. This information is included in the Beacon. The CP identifies the connection by the connection ID (per-flow classification).

2. *Delay and delay jitter:* Isochronous data service provides bounded delay and delay jitter. Bounded delay and delay jitter are achieved using a TDMA based channel access scheme with retransmission option. The CP allocates a fixed transmission duration for each connection.

3. *Reliability:* The service allows the retransmission option to increase the transmission reliability.

4. *Priority:* This service has the highest priority.

In conclusion, the isochronous data service can provide quantitative QoS services.

6.5.2 Priority Asynchronous Data Service

This service, which is designed for applications (i.e., streaming video and audio) that originate at S-nodes, involves the following service characteristics and QoS mechanisms:

1. *Session-oriented service:* An S-node sends the connection request to the CP. The connection request contains a priority value (0-7 value which is similar to the IEEE 802.1D priority definition) or a delay/delay jitter requirement. The CP assigns the session ID (SID) and determines the associated priority based on the priority value or delay/delay jitter requirement information. The classification is based on the SID.

2. *Delay and delay jitter:* Delay and delay jitter are not bounded but are less than the values provided by the asynchronous data service. Priority CSMA/CA channel access is used. Packet scheduling follows a strict priority policy (i.e., a session with a higher priority will receive service first).

3. *Priority:* This service has higher priority than asynchronous data service but lower than isochronous data service.

From the service characteristics presented above, it can be concluded that priority asynchronous data service delivers qualitative (relative) QoS service. The standard does not define traffic policing mechanism. A higher priority session can occupy and starve the bandwidth of lower priority sessions and asynchronous data service.

6.5.3 Asynchronous Data Service

This service is designed for applications (i.e., non-real-time applications) that originate at A-nodes and do not have any specific QoS requirements. This service has the following service characteristics and QoS mechanisms:

1. *Connectionless service:* This service does not require connection setup.
2. *Delay and delay jitter:* Since the channel access mechanism uses CSMA/CA, the delay and delay jitter are variable.
3. *Priority:* This service has the lowest priority.

In summary, asynchronous data service delivers best effort services.

Wireless Metropolitan Area Networks

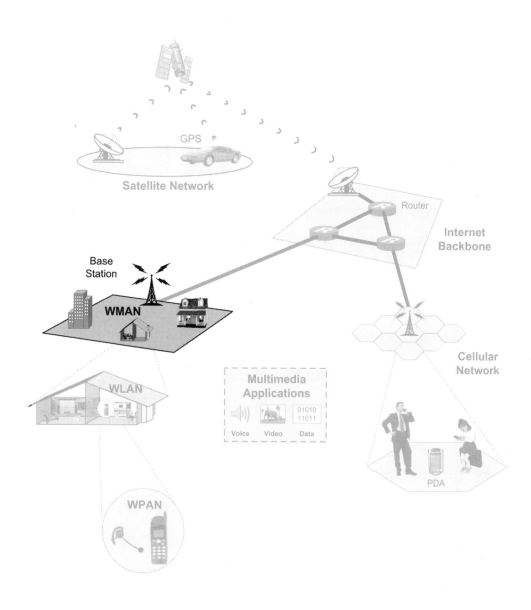

CHAPTER 7

IEEE 802.16

7.1 Introduction

IEEE has been working on the standard for MAC and physical layer specifications for wireless metropolitan area networks (WMANs or WirelessMAN™). The goal of WMANs is to provide high-speed wireless Internet access similar to wired access technologies such as cable modem, digital subscriber line (DSL), Ethernet, and fiber optic. IEEE was motivated by the ability of the wireless technologies to cover large geographic areas without the need to deploy wires.

IEEE published the first standard in 2002. The first approved standard, referred to as IEEE 802.16.1, is geared towards network access between buildings with exterior antennas communicating with a central radio base station (BS) in the 10 to 66 GHz frequency bands. In parallel, the European Telecommunications Standards Institute (ETSI) has been working on a standard that serves similar goals. This effort, referred to as HiperAccess, was completed in 2002, when ETSI decided to adapt IEEE 802.16.1 as its HiperAccess standard.

Since several WMANs can coexist and consequently interfere with each other, the IEEE organization also published in 2001 a recommended practice, referred to as IEEE 802.16.2 for Local and Metropolitan Area Networks, such that mutual interferences between coexisting WMANs are minimized.

IEEE is still working on an addition to the 802.16.1 standard that details MAC and physical layer specifications for the 2 to 11 GHz bands. This effort is referred to as IEEE 802.16a. The European organization, ETSI, has been working on a similar project, referred to as HiperMAN. The organizations are working closely together and it is expected that IEEE 802.16a will be adapted as HiperMAN.

IEEE 802.16 is also working to introduce an amendment, referred to as P802.16c, that will detail System Profiles for the 10 to 66 GHz band. These system profiles help in the deployment of interoperable systems.

The IEEE 802.16 group realized the need for multimedia applications and the required QoS support. Therefore, IEEE 802.16 has included a number of QoS signaling mechanisms. However, the algorithms that use such signaling mechanisms in order to provide QoS support are vendor specific and are left out of the standard. This allows vendors to differentiate their products but still be interoperable.

7.2 IEEE 802.16.1

IEEE 802.16 was designed considering that a metropolitan area can include hundreds or thousands of subscribers which require high-speed connections. The MAC is built to accommodate a point to multipoint topology, reflecting the fact that the BS serves multiple subscriber stations (SSs). The 802.16a MAC will add the capability of serving mesh networks. The MAC addresses the high-speed QoS requirements with a flexible design of uplink (SS to BS) and downlink (BS to SS) channels. The BS has full control on the bandwidth allocation on both channels. Access allocation is provided by the BS via a request-grant mechanism, in which the SS requests access and the BS grants access. The standard realizes that users may have a variety of needs resulting from their legacy voice systems, voice over IP systems, and their TCP/IP packetized systems. It also realizes that, due to different applications, users may execute applications that result in either continuous or bursty traffic, with different QoS requirements.

7.2.1 Architecture

7.2.1.1   Network Topology

The IEEE 802.16.1 is geared toward point to multipoint (PMP) communication topology, in which a BS communicates with numerous stations, or SSs. All connections need to go through the BS (i.e., there is no direct communication between the SSs). Due to the short wavelength of the 10 to 66 GHz frequency bands, with channels of 25 or 28 MHz, the standard requires line of sight between the BS and SS. Data rates in excess of 120 Mbps can be supported. The IEEE 802.16a extension effort supports significantly reduced data rates at 2 to 11 GHz frequency bands. This approach does not require line of sight and supports mesh topology in which connections do not need to go through the BS.

Figure 7.1 shows an example of a network topology in which PMP and mesh topologies are used to cover a large metropolitan area. The in-building communication can be provided through wireless LANs (e.g., IEEE 802.11) or via an implementation of the 802.16a extension. Hence, IEEE 802.16 can provide a complete wireless communication solution for the targeted metropolitan area.

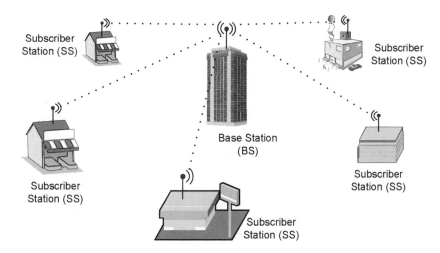

Figure 7.1   Point to Multipoint Network Topology

7.2.1.2   Protocol Stack

The standard is divided into various layers as shown in Figure 7.2. The MAC layer is divided into three sublayers: The Service Specific Convergence Sublayer (CS), the MAC Common Part Sublayer (CPS), and the Security Sublayer.

The Service Specific Convergence Sublayer (CS) transforms incoming network data, received through the CS service access point (SAP) into MAC data packets, preserving or enabling QoS, and allowing bandwidth allocation. This transformation maps external network information into IEEE 802.16 MAC information, such as service flow and connection ID. The current standard details two Convergence Sublayer specifications for ATM CS and for Packet CS. The Packet CS is defined for mapping external packets such as IPv4, IPv6, Ethernet, and virtual local area network (VLAN).

The MAC Common Part Sublayer (CPS) provides access control functionality, bandwidth allocation, connection establishment, and connection maintenance.

The Security Sublayer provides security services including authentication, key exchange, and encryption. Data, physical layer (PHY) control, and management information are exchanged between the MAC CPS and the PHY via the PHY SAP. The PHY layer includes multiple radio options.

7.3 Physical Layer

The IEEE 802.16 defines the physical layer for the 10 to 66 GHz frequency bands assuming line of sight between the BS and SS. It uses burst single-carrier modulation in which all data are sequentially transmitted in a single frequency. The standard allows the use of directional antennas to increase capacity and reduce interference from adjacent transmissions. The physical layer supports wide channel bandwidth of 20 or 25 MHz in the U.S. and 28 MHz in Europe. With 20

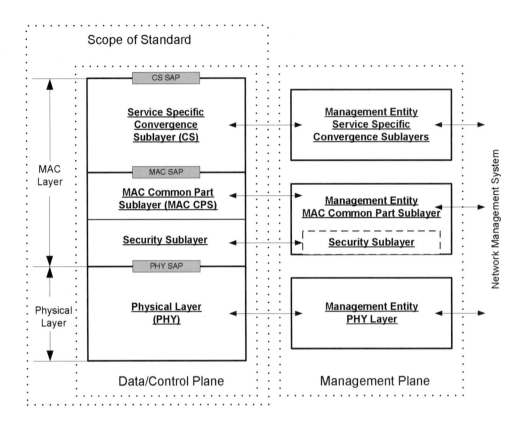

Figure 7.2   IEEE 802.16 Protocol Layer

MHz channel width vendors can support up to 96 Mbps, with 25 MHz up to 120 Mbps, and with 28 MHz up to 134 Mbps. The communication path between an SS and the BS has two directions: uplink, from the SS to the BS, and downlink, from the BS that may reach many SSs. Both uplink and downlink can operate in different frequencies using Frequency Division Duplexing (FDD) or share the same frequencies using Time Division Duplexing (TDD). In both FDD and TDD systems, the uplink and downlink channels are structured into frames. In TDD, the frame is divided into two subframes, uplink and downlink subframes, where the uplink subframe follows the downlink subframe (Figure 7.3). In FDD, the uplink and downlink subframes are concurrent in time but are carried on separate frequencies (Figure 7.4). Each subframe (in TDD or FDD) consists of a number of time slots in which both TDD and FDD systems deploy TDM transmission. All SSs and the BS have to be synchronized and transmit data bursts into predetermined time slots.

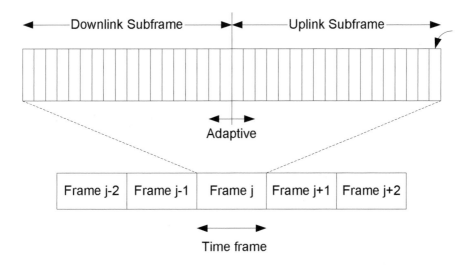

Figure 7.3 TDD Frame Structure

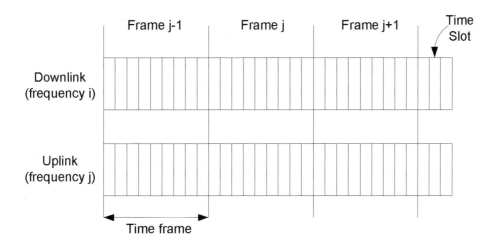

Figure 7.4 FDD Frame Structure

The standard also supports full duplex subscriber stations (which can transmit and receive simultaneously) and half duplex subscriber stations (which can either transmit or receive). For FDD systems, the time slot assignment is different for full or half duplex subscriber stations. When there are half duplex subscriber stations, the BS cannot assign an uplink period for an SS which overlaps with the SS downlink period. For TDD systems, due to the fact that there is no overlapping period between the uplink and downlink channels, both full and half duplex subscriber stations can have the same time slots assignment.

7.3.1   Adaptive Data Burst Profiles

The standard supports adaptive data burst profiling in which transmission parameters, such as modulation and Forward Error Correction (FEC) coding settings, can be modified individually to each SS on a frame-by-frame basis in both uplink and downlink transmissions. The standard allows three types of modulation schemes: Quadrature Phase Shift Keying (QPSK), 16 Quadrature Amplitude Modulation (QAM), and 64 QAM. Different combinations of modulation and FEC provide different transmission robustness and transmission speed (i.e., 64 QAM delivers high transmission speed but is prone to interference; whereas, QPSK delivers low transmission speed but is tolerant to interference). Data burst profiles, which include parameters such as radio modulation type and FEC, are identified by a code called Interval Usage Code (IUC). There are two types of IUC: 1) Downlink Interval Usage Code (DIUC), which identifies the downlink data burst profiles, and 2) Uplink Interval Usage Code (UIUC), which identifies the uplink data burst profiles. DIUC is included in a MAC message called Downlink Channel Descriptor (DCD) message; whereas, UIUC is included in the Uplink Channel Descriptor (UCD) message. DCD and UCD messages are transmitted periodically by the base station in order to define the downlink and uplink channel characteristics, respectively.

7.4   Media Access Control (MAC)

The 802.16.1 MAC is connection oriented. All traffic including inherently connectionless traffic is mapped into a connection. This provides the ability to request bandwidth with its associated QoS and traffic parameters for every connection. Connections are identified by Connection Identifiers (CIDs). The standard allows bandwidth provision to these connections to be either continuous or on demand. Each connection is associated with a service flow (SF) that is a unidirectional flow of MAC packets on a connection that is provided with a particular QoS. The service flow defines the QoS parameters for the packets that are transmitted on this connection. The Service Flow concept is central to the 802.16 MAC since it provides the mechanism for QoS and bandwidth allocation process. The MAC has also some reserved CIDs for purposes such as management, announcements, and broadcasting.

Established connections require active maintenance depending upon the type of service. For example, unchannelized T1 services require virtually no connection maintenance since they have a constant bandwidth allocation. Channelized T1 services require some maintenance due to the dynamic bandwidth requirements. Connections may be also terminated or adjusted based upon users' contracts. The standard supports all of these connection management functions through the use of static configuration and dynamic addition, modification, and deletion of connections.

7.4.1   Channel Access

As described earlier, the communication path has two directions—uplink and downlink. In uplink transmissions (SSs to BS), several SSs share the channel in a TDMA fashion. The Uplink Map Message (UL-MAP) is used to provide the channel access assignment to the subscriber stations. The UL-MAP which is transmitted by the base station at the beginning of the frame defines the uplink channel access as well as the uplink data burst profiles (i.e., UIUC) in the current uplink subframe (see Figure 7.5B-C). The SSs are allowed to transmit data bursts at predetermined time slots as indicated in the UL-MAP. In downlink transmissions (BS to SSs), due to the fact that there is only one station (BS) transmitting the data, the channel access is rather simple. The data packets are transmitted by the BS to all SSs and an SS only picks up the packets destined to it. The Downlink Map Message (DL-MAP) is used to define the downlink data burst profiles (i.e., DIUC) in each time period in the current downlink subframe, in both TDD and FDD systems (see Figure 7.5A).

Both UL-MAP and DL-MAP indicate the starting time slot of each data burst (see Figure 7.6).

7.4.1.1   Downlink Subframe

The standard specifies different downlink frames for FDD and TDD systems.

The FDD downlink subframe (Figure 7.7) starts with a preamble, DL-MAP, and UL-MAP. The preamble helps in the physical layer transition and synchronization. The DL-MAP defines bursts' start times on the downlink for both TDM and time division multiple access (TDMA) methods. The downlink subframe is divided into TDM and TDMA portions, where the TDMA portions follow the TDM portions. The TDM portions are not separated by gaps or preambles since there is only one transmitter, the BS. The TDM portions contain data transmitted to one or more of the following: full duplex SSs, half duplex SSs scheduled to transmit later in the frame, and half duplex SSs not scheduled to transmit in this frame. On the other hand, the TDMA portions are separated by preambles and gaps. The TDMA portion is used to transmit data to any half duplex SSs scheduled to transmit earlier in the frame than they receive data. This allows an individual SS to decode a specific portion of the downlink without the need to decode the entire downlink subframe.

The TDD downlink subframe (Figure 7.8) starts with a preamble, DL-MAP, and UL-MAP. The preamble helps in the physical layer transition and synchronization. The TDD downlink subframe contains data packets that are transmitted to SSs and ends with a Transmit/Receive (Tx/Rx) time gap. In the TDD downlink subframe there are only TDM portions.

7.4.1.2   Uplink Subframe

The uplink subframe (Figure 7.9) is used by the SSs to transmit to the BS.

The uplink subframe includes the following three periods: Initial Maintenance period, Request Contention Opportunities period, and Scheduled Data grants period. The different periods are identified by their respective UIUC that is specific to an uplink period. The BS announces these periods and associated burst classes in the preceding downlink subframe's UL-MAP. The BS can specify such periods in any order and length. The BS can group the periods allocated to the Initial Maintenance and Request Contention Opportunities and leave the remaining periods for data transmission. We provide next a more detailed description of each period.

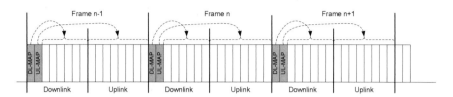

(A) TDD (Minimum Time Relevance)

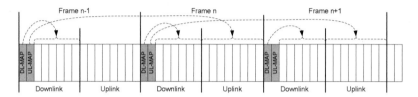

(B) TDD (Maximum Time Relevance)

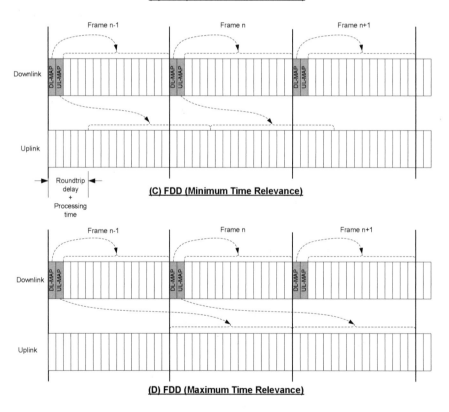

(C) FDD (Minimum Time Relevance)

(D) FDD (Maximum Time Relevance)

Figure 7.5   MAC Control Information

UL-MAP and DL-MAP indicate the starting time slot of each data burst

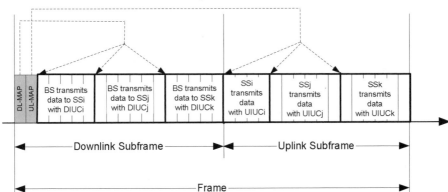

Figure 7.6 DL-MAP and UL-MAP

In the Initial Maintenance period, identified by UIUC = 2, the SS sends transmissions or bursts needed to carry out Initial Maintenance functions. For example, ranging requests are sent by the SS at the initialization phase and periodically thereafter. The BS uses such requests to determine network delay and request power or downlink burst profile changes. In this period also new stations may join the network. Since several SSs can access the channel simultaneously, collisions can occur in this period.

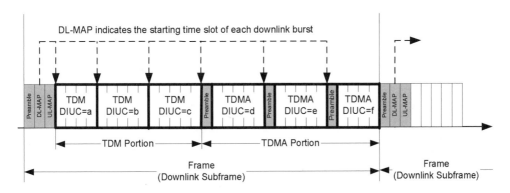

Figure 7.7 FDD Downlink Subframe Structure

In the Request Contention Opportunities period, identified by UIUC = 1, the SSs request bandwidth based upon multicast and broadcast polls by the BS. Since several SSs can access the channel simultaneously, collisions can occur in this period.

In the Scheduled Data grants period, identified by UIUC that defers from 1 or 2, the SS transmits data based on grants allocated by the BS. An SS transition gap separates each Scheduled Data period to allow transition from one period with a specific burst profile to another period which may have a different burst profile. Each such period starts with a preamble to allow the new SS to synchronize.

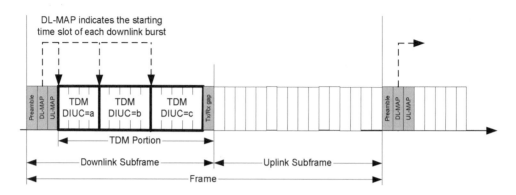

Figure 7.8 TDD Downlink Subframe Structure

7.4.1.3   Contention Resolution

Since several SSs can access the channel simultaneously during the Initial Maintenance and Request periods, their respective transmissions can collide. If such a collision occurs the transmitting SSs will go through a contention resolution process. It is possible that the SS will have multiple uplink service flows, each associated with its CID. The SS will consider these CIDs or their QoS needs in this process. The mandatory process by the standard is based on a truncated binary exponential backoff, with an initial backoff window and a maximum backoff window controlled by the BS. The algorithm is defined in the standard.

7.4.2   Bandwidth Request Mechanisms and Bandwidth Allocation

The standard defines various mechanisms for the SS to access the shared uplink and request transmission opportunities (bandwidth) and for the BS to grant such transmission opportunities. The key mechanisms of the request-grant process are the bandwidth request mechanism and the bandwidth allocation. These mechanisms enable multiple types of service flows which support a wide range of applications. First, we introduce the bandwidth request and the bandwidth allocation mechanisms and then we describe each type of service flow offered by the standard.

7.4.2.1   Bandwidth Request Mechanism

Bandwidth is always requested on a connection (i.e., CID) basis. Each connection in an SS requests bandwidth through an MAC message called BW Request message. BW Request can be sent in a stand-alone packet or piggyback on another packet. The request is based on the number of bytes needed to carry the MAC packet without considering the physical layer overhead. The requested bandwidth can be either incremental, meaning how much additional bandwidth is needed, or aggregate, meaning how much total bandwidth is needed. When the BS receives an incremental bandwidth Request, it adds the quantity of bandwidth requested to its current allocated bandwidth. When the BS receives an aggregate bandwidth Request, it replaces the allo-

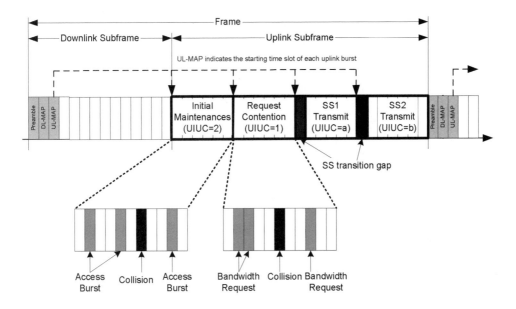

Figure 7.9 Uplink Subframe Structure

cated bandwidth with the quantity of bandwidth requested. With piggyback BW Request, the connection can only request incremental bandwidth. With stand-alone BW Request, the connection can request either incremental or aggregate bandwidth. BW Request can be initiated directly by the connection or initiated in response to receiving a Polling Message from BS. BW Request can be transmitted during the uplink subframe in two different ways: 1) Contention request opportunities in which BW Request is transmitted in Request Contention period and 2) Contention-Free Request Opportunities in which BW Request is transmitted in the predefined time slot of uplink subframe indicated by Information Element (IE) of UL-MAP or is piggy-backed with other packets.

The polling mechanism is geared for detailed and flexible requests but requires more overhead. BS issues unicast polling packets for the SS as well as multicast and broadcast polls. The polling process does not include an explicit packet from the BS to the SS. Rather, the SS is allocated bandwidth for its potential request in the downlink subframe field that details the coming uplink transmission opportunities, referred to as uplink map.

The standard also allows the BS to poll stations in groups or as a whole. This mode can be used in cases where the BS decides that there is not enough bandwidth to support individual unicast polls or in cases where there are many inactive SSs. Certain connection CIDs are reserved for such multicast and broadcast polls. The SSs addressed in this group polling will need to contend during the uplink period to send their transmission requests.

7.4.2.2   Bandwidth Allocation

As mentioned in the previous section, each SS requests bandwidth on a connection basis. The BS grants or allocates bandwidth by deploying one of the following two modes: Grant Per Subscriber Station (GPSS) and Grant Per Connection (GPC). In GPC mode, the BS allocates bandwidth to individual connections (CID). In GPSS mode, the BS allocates bandwidth for individual SSs. In this case, the SS is responsible for distributing this allocated bandwidth to each connection within the SS. The SS decides which traffic should be transmitted first, considering each application's traffic needs and QoS requirements. Thus, GPSS mode requires more intelligence in the SS than does the GPC mode. This allows vendors to design different SS configurations. A vendor may choose a GPC design for simplicity. If the 10 to 66 GHz frequencies are used, the standard allows only SSs with GPSS mode.

The BS allocates bandwidth to a connection or a station by granting the transmission opportunity in the UL-MAP. The bandwidth allocation is mainly based on the following factors:

- The amount of bandwidth requested by the connections
- QoS parameters of delay and bandwidth needed by the current applications at the SS as reflected in the service flow parameters
- Available network resources

The BS can grant transmission opportunities for multiple connections to a single SS, each with a different QoS level. The BS may not grant bandwidth to the requesting SS due to reasons such as: the bandwidth requested by the SS is not available, the BS received a request that included errors, and the request did not go through due to collisions with other SSs or due to errors in the wireless media.

7.4.3   Service Flow Scheduling Service

The standard defines four types of service flows that provide QoS support for a wide range of applications. The services include: Unsolicited Grant Service (UGS), Real-Time Polling Service (rtPS), Non-Real-Time Polling Service (nrtPS), and Best Effort (BE) Service.

7.4.3.1   Unsolicited Grant Service (UGS)

The Unsolicited Grant Service (UGS) supports real-time service flows that generate fixed size periodic data packets, such as T1. This service provides periodic, fixed size grants that avoid the overhead and latency of frequent SS redundant requests. It also ensures that grants are available to meet the continuous needs of the service flow.

A connection with UGS service flow is prohibited from using any contention request opportunities. The BS will not provide any unicast request opportunities for this connection. The Information Element (IE) of the UGS service flow includes parameters such as the UGS size, nominal grant interval, the tolerated grant jitter, and the request/transmission policy. The SS updates the BS regarding the state of the UGS service flow, using special fields called Slip Indicator (SI) flag in the packet header. The SI flag indicates excessive queue length due to conditions such as a lost allocation map or due to clock mismatch between the SS and the BS. In such cases the BS may grant additional bandwidth to compensate for bandwidth insufficiency. An SS that is granted UGS opportunity cannot "steal" bandwidth. In other words, the SS cannot use a portion of the allocated bandwidth to send another bandwidth request rather than sending data. Also it can not piggyback a request in the transmitted data. However, the SS can indicate that it needs a poll for non-UGS service by setting a special field, referred to as Poll-Me bit, in the packet header. This mechanism saves bandwidth, as the BS can poll such stations only if the Poll-Me bit is set. Once the BS receives this indication, it may follow with individual polling such as rtPS and nrtPS.

7.4.3.2   Real-Time Polling Service (rtPS)

The BS polls SSs with the purpose of allocating to these SSs bandwidth specifically for the purpose of making bandwidth requests. The rtPS supports real-time service flows that have periodic data packets of various sizes, such as MPEG streams. This service provides periodic, unicast request opportunities, to respond to the needs of the service flows. It allows the SS to dynamically specify the size of the requested grant. Compared to UGS, this service has additional overhead due to the polling process. However, it can handle variable grant sizes compared to UGS that can handle only fixed grant sizes. The standard specifies that the BS needs to poll the SS periodically in order to allow it to send unicast request opportunities. The polling frequency (the number of polls in a certain period of time) is not specified in the standard. In common practice, the BS should poll the SS frequently enough to meet the delay and bandwidth requirements of real-time applications. In this case, the SS is prohibited from using any contention request opportunities in order to avoid unpredictable delays. A connection with rtPS service flow is allowed to "steal" bandwidth if it is in the GPSS mode. rtPS service flow is also allowed to piggyback a BW Request on a data packet.

7.4.3.3   Non-Real-Time Polling Service (nrtPS)

The Non-Real-Time Polling Service (nrtPS) supports non-real-time service flows that have data packets of various sizes, such as FTP. The service provides consistent unicast request opportunities even during network congestion. The standard specifies that a BS needs to poll the SS on a regular basis (periodically or nonperiodically) and allow it to send unicast request opportunities. In this case, the SS is allowed to use contention request opportunities for the connection. In other words, the SS can simultaneously send unicast request packets in response to the BS Poll and content for request opportunities. A connection with nrtPS service flow is allowed to "steal" bandwidth if it is in GPSS mode. nrtPS service flow is allowed to piggyback a BW Request on a data packet.

7.4.3.4   Best Effort (BE) Service

The Best Effort (BE) Service supports service flows that do not require QoS support. The SS issues its requests in a contention period. An SS that is granted a request via BE is allowed to "steal" bandwidth if it is in GPSS mode. BE Service flow is allowed to piggyback a BW Request on a data packet.

The standard details extensive signaling techniques and access mechanisms for each service flow, but the details of bandwidth allocation, scheduling, and reservation management intelligence are out of the standard's scope and are left to be vendor specific. The ability to employ different combinations of these access mechanisms allows vendors to differentiate their products, tailor solutions to unique needs and users, optimize system performance, and use different bandwidth allocation algorithms while maintaining consistent interoperability. For example, contention may be activated to avoid polling of SSs that have been inactive for a long period of time.

7.4.4   Network Entry and Initialization

An SS wishing to enter the network has to go through the following steps:

1. Scan for a downlink channel and establish synchronization with the BS. The SS can either recall the downlink channel from its memory or continuously scan possible channels of the downlink frequency band until it finds a valid downlink signal. After it finds a downlink channel it needs to synchronize and attempt to acquire the channel control parameters for the downlink by searching for the DL-MAP sent by the BS.
2. Obtain transmit parameters. The SS searches for an uplink Channel Descriptor message from the BS in order to retrieve the transmission parameters for a possible uplink channel.
3. Adjust local parameters (e.g., transmit power) based on approved values or messages from the BS.
4. Negotiate basic capabilities. The SS informs the BS of its basic capabilities by sending an appropriate message to the BS. The BS acknowledges this message.
5. Authorize SS and perform key exchange. The BS goes through a process of authorizing the SS to enter the network and exchanges security keys with the SS.
6. Perform registration. The BS sends additional management messages and the SS becomes managed by the BS.
7. Establish IP connectivity. The SS receives an IP address from the BS.
8. Establish time of day. The SS and BS need to have equal timing information.
9. Transfer operational parameters. The BS sends additional configuration information to the SS.

7.5 QoS Support

The principal mechanism of IEEE 802.16 standard for providing QoS support is to associate a packet with a service flow. A service flow is a unidirectional flow of packets that provides a particular QoS. The standard details the mechanisms of how to allocate bandwidth and how to send the BW Requests in each service flow as described in the previous section. To summarize, the service flows and the supported applications are listed below:

- UGS Service flow supports real-time applications with constant bit rate (CBR) such as voice over IP and circuit emulation.
- rtPS Service flow supports real-time applications with variable bit rate (VBR) such as streaming video and audio.
- nrtPS Service flow supports non-real-time applications which require better service than BE, such as high-bandwidth FTP.
- BE Service flow supports applications that do not have any QoS requirements.

Each network application, first, has to register with the network. The network will associate the application with a service flow by assigning an unique Service Flow ID (SFID). All packets must be tagged with this assigned SFID in order for the network to provide the appropriate QoS. When the application wants to send data packets, it is required to establish a connection with the network and receives a unique CID assigned by the network. Therefore, the IEEE 802.16 data packets include both CID and SFID.

Next we describe QoS support from the following two aspects: QoS provision and QoS mechanisms.

7.5.1   QoS Provision

IEEE 802.16 broadband wireless access standard aims to provide fixed wireless access between the subscriber station (residential or business customers) and the Internet Service Provider (ISP) through the BS. The service model involves the following steps:

- In order to receive network service, the customers have to subscribe or register with the ISP or another entity that has the authority to control the network.
- Whenever the customers want to use the service (i.e., transmit data), they need to establish the connection with the network.
- If customers require QoS support for their applications, they are required to include a QoS parameter set that can be defined in any of the following ways: explicitly specifying all traffic parameters, indirectly referring to a set of traffic parameters via specifying a Service Class Name, or specifying a Service Class Name along with modifying the parameters.
- Considering this QoS parameter set and the available network resources, the network decides if it can provide the appropriate QoS support.

The standard's set of tools that support QoS for both uplink and downlink traffic include: configuration and registration functions for service flows, a signaling function for dynamically establishing QoS based service flows and traffic parameters, scheduling and QoS traffic parameters for uplink and downlink service flows, grouping of service flow properties into service classes to allow grouping of requests.

The standard's QoS provision, based on the "envelopes" model shown in Figure 7.10, defines several sets of QoS parameters:

- *ProvisionedQoSParamSet:* A set of external QoS parameters provided to the MAC, for example, by the network management system.
- *AdmittedQoSParamSet:* A set of QoS parameters for which the BS and possibly the SS are reserving resources since the associated service flows have been admitted by the BS. Resources are not limited to bandwidth and can include resources such as memory.
- *ActiveQoSParamSet:* A set of QoS parameters that reflect the actual service being provided to the associated active service flows.

The standard assumes an Authorization Module in the BS that approves or denies every change of QoS parameters associated with a service flow. This module can provision a service flow immediately or can defer the service flow activation to a later period.

The standard presents an "envelope" model that limits the possible values of the AdmittedQoSParamSet and ActiveQoSParamSet. The standard recognizes two models: Provisioned Authorization Model and Dynamic Authorization Model. In the Provisioned Authorization Model the parameters are provided, for example, by the network management system. In the Dynamic Authorization Model, the authorization module issues its decisions based on its vendor specific implementation. This implementation may use routines that communicate to external policy servers for additional information and input. The relationships among the QoS Parameter Sets in each model are shown in Figure 7.10. The ActiveQoSParamSet is always a subset of the AdmittedQoSParamSet, which is always a subset of the authorized "envelope." In the Dynamic Authorization Model, this envelope is determined by the Authorization Module (labeled as the AuthorizedQoSParamSet). In the Provisioned Authorization Model, this envelope is determined by the ProvisionedQoSParamSet.

7.5.2   QoS Mechanisms

In this section, we revisit the service flow and its mechanisms, emphasizing the QoS aspects.

7.5.2.1   Classification

All packets generated by active applications are tagged with CID and SFID. The classification module identifies the packets based on these tags and forwards them to the corresponding queues (see Figure 7.11).

Individual applications can establish individual connections. IEEE 802.16 can provide per-flow classification which supports per-flow QoS services.

7.5.2.2   Channel Access

As described in Section 7.5.1, the channel access method uses TDM for the downlink and TDMA for the uplink. The channel access is controlled by the BS through the UL-MAP and DL-MAP. Both TDM and TDMA are collision-free channel access schemes which provide tight channel access control. Therefore, these schemes can provide QoS support for applications with strict QoS requirements.

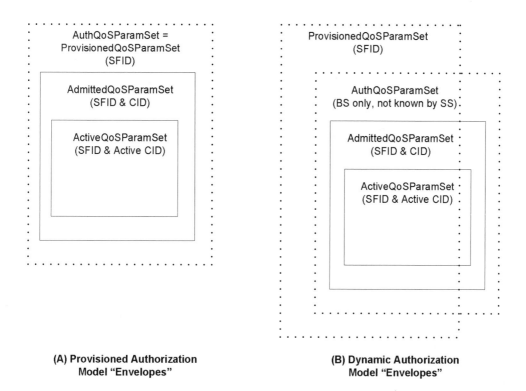

**(A) Provisioned Authorization
Model "Envelopes"**

**(B) Dynamic Authorization
Model "Envelopes"**

Figure 7.10 QoS Provision Model

7.5.2.3   Packet Scheduling

The packet scheduling module allocates bandwidth for connections in terms of the number of time slots allocated per connection on the TDM channel. This module also determines when a connection is allowed to transmit the data. This information is defined in the UL-MAP and DL-MAP. There are two packet scheduling modules: uplink packet scheduling and downlink packet scheduling. The downlink packet scheduling module is relatively simple because all the queues reside in the BS. Therefore, the downlink packet scheduling module can easily retrieve the state of the queues. The standard does not define the downlink packet scheduling algorithm. The downlink packet scheduling algorithms can be similar to the scheduling algorithms used in a router. The uplink packet scheduling module is more complex because the queues are distributed among multiple SSs. The uplink packet scheduling module obtains the state of the queues and the bandwidth requirements from each connection through the BW Request message as defined in the standard. The uplink packet scheduling determined by the uplink packet scheduling algorithm is reflected in the UL-MAP. Figure 7.12 illustrates the uplink scheduling process.

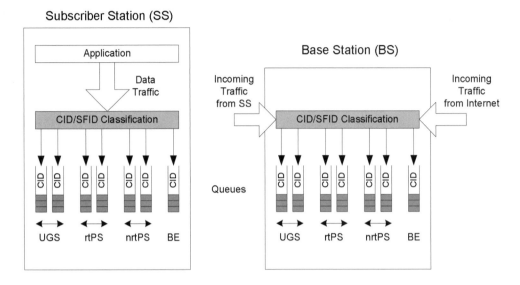

Figure 7.11 Classification Module

The standard does not define the scheduling algorithm that determines the UL-MAP. The uplink scheduling algorithm has two modes of allocating bandwidth, GPC and GPSS:

- In GPC mode, the uplink scheduling algorithm distributes the bandwidth on a per-connection basis. The intelligence of the uplink scheduling algorithm resides in the BS. The packet scheduling module at each SS is simple because it just follows the UL-MAP.
- In GPSS, the uplink packet scheduling distributes bandwidth on a per-station basis. The SS distributes its allocated bandwidth to each connection within the SS. Compared to GPC, the GPSS based uplink packet scheduling algorithm that resides at the BS is less complex. On the other hand, the packet scheduling algorithm that resides at each SS is more complex.

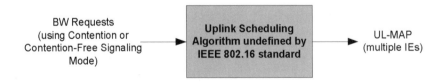

Figure 7.12 Uplink Packet Scheduling Process

Figure 7.13 summarizes IEEE 802.16 QoS architecture which can deliver multiple levels of QoS service: Quantitative Service, Qualitative Service, and BE Service.

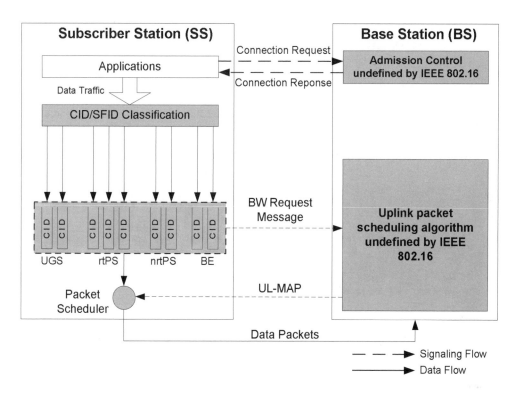

Figure 7.13 IEEE 802.16 QoS Architecture

7.6 IEEE 802.16a

The IEEE 802.16a group is considering an extension to the standard. This section provides a glimpse into the thoughts and ideas discussed in the group. The final extension, when approved, may be significantly different than these thoughts.

IEEE 802.16a focuses on extending the standard support to a mesh network topology in which direct line of sight is not required and in which SSs can communicate with each other (see Figure 7.14). This mesh topology is proposed in addition to the point to multi-point topology covered in the approved standard. Another addition is extending the physical layer to cover frequencies between 2 and 11 GHz in addition to the 10 to 66 GHz covered in the approved standard. The MAC includes the addition of an ARQ (Automatic Repeat Request) mechanism for handling communication errors as well as further signaling support for traffic scheduling algorithms that provide QoS support.

A mesh network topology is a much more complex architecture than the point to multi-point topology which extends the geographical coverage of the network. In a mesh topology, each SS can communicate with another SS, and each SS can forward traffic transmitted between the BS and the destination SS. The support of a mesh topology is more complex since we need to consider issues such as routing, collision avoidance and resolution between adjacent SSs, synchronization, and more.

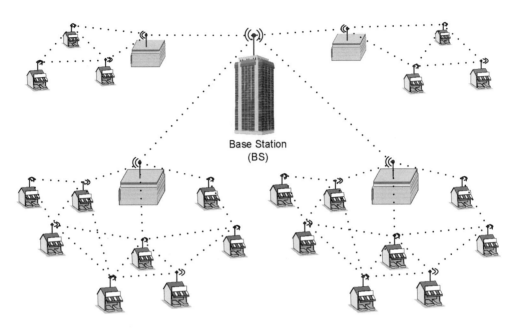

Figure 7.14 IEEE 802.16a Mesh Network Topology

The 2 to 11 GHz band considered by IEEE 802.16 provides a physical environment in which line of sight is not required. This is due to the longer wavelength as compared with the 10 to 66 GHz physical environment specified in the approved standard. The channel bandwidths used in this proposed extension are between 1.5 and 28 MHz. This addition proposes three radio technologies: single-carrier modulation format, OFDM (Orthogonal Frequency Division Multiplexing), and OFDMA (Orthogonal Frequency Division Multiple Access). OFDMA is a version of OFDM in which transmissions are carried in simultaneous subchannels compared to one channel in OFDM. OFDMA allows for increased bandwidth as well as increased resiliency to errors.

The proposal details communication in the 2 to 11 GHz frequencies where there are two license-free bands, 2.4 GHz and 5.4 GHz, that were allocated primarily for WLAN users. To avoid interfering with such users, referred to as Primary Users, the proposal defines algorithms to detect and avoid interfering with WLAN users. Referred to as Dynamic Frequency Selection (DFS), this process is an ability of a system to switch to different physical RF channels between transmit and receive activity based on channel measurement criteria. The DFS procedures test channels for primary users discontinue communication once such primary users are detected, and periodically schedule for channel testing to detect new primary users.

7.6.1   IEEE 802.16a MAC Discussions

The proposal for the mesh topology suggests both distributed and centralized scheduling mechanisms. Each mechanism can be employed with directional antennas, adaptive antenna systems, or regular omni-directional antennas. In the mesh topology, SSs that join the network do not need to interact directly with the BS but rather with their neighbor SSs.

In distributed scheduling, all the SSs need to coordinate their transmissions with their neighbors that are up to two hops away. They broadcast their available resources, requests, and grants to all of their neighbors. They can also schedule transmissions by using directed requests and grants. The proposal does not differentiate in the scheduling mechanism between downstream and upstream channels. However, SSs need to ensure that their resulting transmissions do not collide with already scheduled data and control traffic. Distributed scheduling can be executed in either a coordinated or an uncoordinated manner. The coordinated scheme employs scheduling packets that are transmitted in collision-free, regular periods within scheduling control subframes. The uncoordinated scheme allows contention based access while avoiding conflicts with the schedules established using the coordinated scheme. In case of collisions, the SS backs off and tries to retransmit.

In centralized scheduling the BS allocates all resources. The BS collects the resource requests from all the SSs that are up to a predetermined hop count from this BS. Using this information, the BS grants resources for each communication link for both downstream and upstream channels. The BS announces its decision to all of the SSs within its hop range. SSs that receive either requests from other SSs or grants from the BS forward these messages to other neighbor SSs such that all participants within the mesh are covered. To avoid multiple redundant transmissions of the same data, the mechanism assumes the existence of an algorithm where SSs know their hop count from the BS. The proposal suggests that the grant messages will not contain the actual granted schedule but rather parameters from which each SS can compute its allocated resources using a predetermined algorithm.

Wireless Personal Area Networks

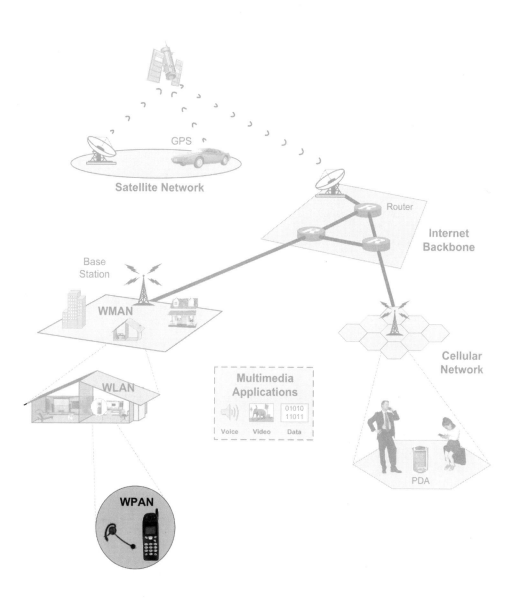

CHAPTER 8

Bluetooth

8.1 Introduction

The Bluetooth standard for wireless personal area networks is defined by the Bluetooth Special Interest Group (SIG) originally founded by Ericsson, IBM, Intel, Nokia, and Toshiba. Presently, Bluetooth SIG also includes companies such as 3Com, Agere, Microsoft, Motorola, and hundreds of other member companies. Bluetooth is named after Denmark's first king Harald Blaatand (Bluetooth in English) (940-981), son of Gorm the Old King of Denmark and Thyra Danebod. Some suggest that the king liked eating blueberries and thus his teeth became stained with the blue color. Harald Blaatand united Denmark and Norway. This inspired the name "Bluetooth," meaning "uniting" devices through Bluetooth. Bluetooth was first introduced by Ericsson as a practical technology for small form factor, inexpensive, short-range radio links between PCs, handheld devices, mobile phones, and other computing and electronic devices that communicate and interface with cellular phones and the Internet. Many companies have already introduced Bluetooth compliant products.

First we would like to compare Bluetooth with infrared technology. The Infrared Data Association (IrDA) has introduced two standards: IrDA-Data, for high-speed, short-range, line-of-sight, and point-to-point data transfer, and IrDA-Control, for lower speed communication such as wireless keyboards and joysticks. The range of IrDA is around 1 meter and requires line of sight between the transmitter and receiver. Moreover, IrDA communication is subjected to light interferences. The biggest advantage of IrDA over Bluetooth for point-to-point communication is its high throughput, which makes it suitable for multimedia applications requiring high-speed transmissions. However, Bluetooth has larger range, does not require line of sight, and is not prone to light interferences.

Bluetooth 1.1 specifies a wireless technology in the 2.4 GHz Industrial, Scientific, and Medical (ISM) band that supports a range of 10 meters. Operating on the unlicensed ISM band,

Bluetooth requires the use of spread spectrum transmission technology which is resistant to interference. Bluetooth deploys fast rate frequency hopping (compared to other Frequency Hopping Spread Spectrum (FHSS) wireless technologies) which provides robust data transmission. It supports simultaneous transmissions of both voice and data. Up to eight data devices can be connected in a cell, referred to as *piconet*. Up to 10 cells or piconets can exist within a 10 meter range. These cells are arranged in a multi-cell, referred to as *scatternet,* which increases Bluetooth's coverage range to around 100 meters. Each piconet can support up to three simultaneous full duplex voice devices. Bluetooth 1.1 standard uses a combination of circuit and packet switching. It can support an asynchronous data channel, up to three simultaneous synchronous voice channels, or a channel which simultaneously supports asynchronous data and synchronous voice. Each voice channel supports a 64 kbps synchronous voice channel in each direction. The asynchronous channel can support four asymmetric channels with data rates of up to 723 kbps and 57 kbps in the return direction, or 434 kbps for a symmetric channel.

The earlier Bluetooth 1.0 version had problems related to interoperability. For example, Bluetooth uses an FHSS radio technology in 2.4 GHz frequency band with 79-hop frequency hopping sequences. To maintain communication, the stations have to synchronize on the same hopping sequence. Some European countries allowed only 23-hop frequency hopping sequences, which created an interoperability issue. The newer Bluetooth 1.1 specification addressed this issue as well as many other interoperability issues.

Future Bluetooth 2.0 standard discusses higher speeds, improved functionality, different radios, support for ad hoc peer-to-peer networks, support for devices not currently included such as keyboards, support for high-end multimedia such as video and music, and more.

8.2 Architecture

8.2.1   Network Topology

Bluetooth defines three topologies: point-to-point (Figure 8.1), single cell (piconet) (Figure 8.2) and multi-cell (scatternet) (Figure 8.3). In a point-to-point architecture two devices communicate directly while one device becomes the master and the other becomes the slave. In a piconet, up to eight active devices communicate while one device becomes the master and the rest become slaves. Many more devices can be in "parked" mode, meaning non-active and occasionally listening to the master for synchronization and broadcast messages. The master defines the frequency hopping sequence and the clock for all the devices in the piconet. A scatternet can

Master Node   Slave Node

Figure 8.1   Point-to-Point Topology (single slave operation)

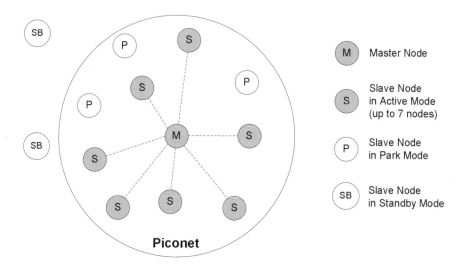

Figure 8.2 Piconet

connect up to 10 piconets. The connection between the piconets is handled by a device (either master or slave) that participates in both piconets. For example, the device can serve as a slave in both piconets or the device can serve as the master in one piconet as well as a slave in another piconet. To avoid collisions and interference between piconets, each piconet uses a different frequency hopping series which is uniquely defined by the master. However, the more piconets are active in the same area, the higher the chance that such piconets will transmit on the same frequency, leading to more collisions and interference.

8.2.2   Protocol Stack

As shown in Figure 8.4, the standard details various protocol layers: the Bluetooth Radio, the Bluetooth Baseband, the Link Manager (LM), the Logical Link Control and Adaptation Protocol (L2CAP), the Host Control Interface (HCI), the Service Discovery Protocol (SDP), the Audio/Telephony Control protocol Specification (TCS), the Radio Frequency oriented emulation of the serial communication ports (RFCOMM), the Human Interface Device (HID), TCP/IP, and other high-level protocols.

The Bluetooth Radio layer, which is parallel to the physical layer in the seven-layer Open Systems Interconnection (OSI) model, defines the requirements for a 2.4 GHz Bluetooth transceiver (i.e., frequency band, radio parameters, transmitter and receiver characteristics). The Baseband layer describes the specifications of the Bluetooth Link Controller which executes the baseband protocols and other low-level link functions. The Baseband layer includes the following modules: LM, HCI, and L2CAP. The LM protocol specifies the link setup and control. The HCI provides a command interface to the Baseband Link Controller and LM and access to the hardware status and control registers. The L2CAP supports higher level protocol multiplexing,

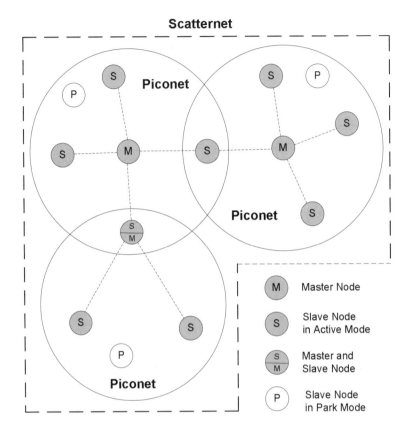

Figure 8.3 Scatternet

packet segmentation, and packet reassembly, and it conveys QoS information. The RFCOMM protocol provides emulation of serial ports over the L2CAP protocol. The Service Discovery Protocol (SDP) provides ways for applications to discover which services are available.

8.3 Physical Layer

The Bluetooth radio uses FHSS in 79 frequencies, each in a 1 MHz range from 2.402 GHz to 2.480 GHz. The radio transceiver changes the hopping frequency in a pseudorandom fashion determined by the master. The hopping rate is 1600 hops per second. GFSK (Gaussian Frequency Shift Keying) is used for the modulation scheme. Each device is classified into power classes 1, 2, and 3. Power class 1 is for long-range transmission of around 100 meters. Power class 2 is for short-range transmission of around 10 meters with less power output than power class 1. Power class 3 is for short-range transmission of around 10 cm, using minimal transmission power. The power control function regulates the transmission power for optimizing the power consumption and channel interference level.

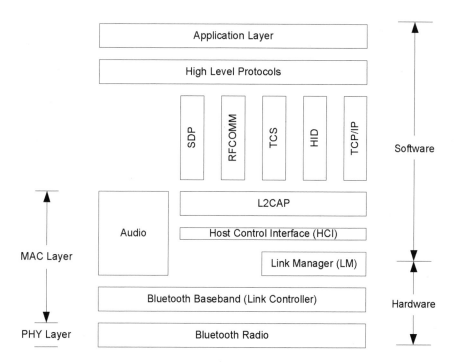

Figure 8.4   Bluetooth Protocol Stack

8.4   Bluetooth Baseband

The Baseband executes the key functions of the Media Access Control (MAC) layer and Data Link Control (DLC) layer. The MAC functions include channel access and packet handling through asynchronous and synchronous links. The link control protocol of the Baseband is implemented as a Link Controller, which works with the LM for carrying out link level routines such as link connection and power control.

8.4.1   Channel Definition and Channel Access

A piconet consists of a master node, active slave nodes (up to seven nodes), and several inactive slave nodes (up to 200 nodes). All connections (or links) in the piconet take place only between the master and slaves in point-to-point or point-to-multipoint fashion. There is no direct connection between slaves. The communication channel is structured into time slots, each with duration of 625 μs. The master uniquely defines the frequency hopping sequence based on its device address (48 bits). Generally, all devices hop or change the frequency every time slot (with max hopping rate of 1600 hops per second, which corresponds to 625 μs time slot duration) except in the case of multi-slot packet transmission, which we will describe later in this section. All slaves (both active and inactive) in a piconet are required to synchronize the clock and frequency hopping sequence with the master. A packet transmission starts at the beginning of a time slot and can last for one, three, or five time slots.

Bluetooth deploys a full duplex, Time Division Duplex (TDD) protocol in which transmission and reception alternate. By alternating time slots, the master and slave can transmit packets according to the TDD protocol. The master starts its transmission only in even numbered time slots and the slave starts its transmission only in odd numbered time slots. Figure 8.5 illustrates a single time slot packet transmission.

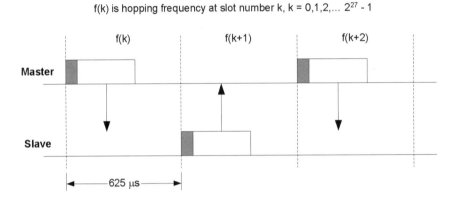

f(k) is hopping frequency at slot number k, k = 0,1,2,... 2^{27} - 1

Figure 8.5 Example of a Single Time Slot Transmission Using TDD

A single-slot packet is transmitted with a hopping frequency of the current Bluetooth clock value. In case of a multi-slot packet, the packet will be transmitted with the hopping frequency of the first time slot for the entire duration of packet transmission. The multiple time slot transmission is shown in Figure 8.6.

Bluetooth aims to support both voice and data transmission by using a combination of both circuit and packet switching protocols. Bluetooth Baseband defines two types of transmission links between a master and a slave: synchronous connection-oriented (SCO) link and asynchronous connectionless (ACL) link. SCO is used for synchronous packet transmission while ACL is used for asynchronous data transmission.

8.4.1.1   SCO and ACL links

The SCO link is a symmetric, point-to-point link between the master and a single slave in the piconet. The SCO link is established by the master by sending an SCO setup packet via the LM protocol. This packet contains timing parameters such as reserved time slots. The master maintains the SCO link by reserving time slots at regular intervals—that is, establishing a circuit switched connection. A slave can support up to three SCO links from the same master or two SCO links if the links originate from different masters. The master can support up to three simultaneous SCO links to the same slave or to different slaves. With the reserved time slots at regular intervals, the SCO link provides bounded delay channel access which is suitable for time sensitive applications such as voice. The master sends packets, referred as SCO packets, at regu-

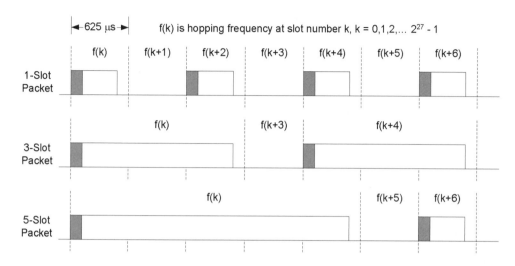

Figure 8.6   Multi-Slot Packets

lar intervals, called SCO intervals and counted in slots in the reserved master-to-slave slots. The
SCO slave is allowed to respond with an SCO packet in the following reserved slave-to-master
slot. Figure 8.7 shows two SCO links with different SCO intervals. Bandwidth allocation to
SCO links and channel access delay are determined by the SCO interval.

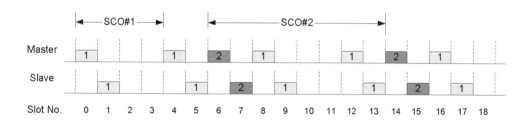

Figure 8.7   SCO Links

There is no packet retransmission in SCO transmissions. The error handling relies on the forward error correcting (FEC) scheme and the codec used. SCO is designed to be used by voice
packet transmissions. Bluetooth specification categorizes the voice packet into three types: HV1,
HV2, and HV3. HV stands for high-quality voice. The differences between the three types of
HV are the number of information bytes and the error correcting scheme. HV1 provides the lowest number of information bytes (low data rate) but uses the most robust error correction scheme
(i.e., 1/3 FEC). HV3 provides the highest number of information bytes (high data rate) but does
not use any error correction scheme. While achieving high data rate, HV3 is susceptible to channel interference. Error handling for HV3 only relies on the codec of voice content.

The ACL link is a point-to-multipoint link between the master and all the slaves in the piconet. In the slots not reserved for SCO packets, the master can establish an ACL link on a per-slot basis (i.e., packet switched connection) to any slaves, including the slave(s) already engaged in an SCO link. Only a single ACL link between a master and a slave can exist. That means that multiple packet switched connections on a slave are aggregated to the same single ACL link. ACL supports both asynchronous and isochronous services. For most ACL packets, packet retransmission is supported to ensure integrity. A slave can respond to an ACL packet in the slave-to-master slot only if it has been addressed in the preceding master-to-slave slot. ACL packets that are not addressed to a specific slave are considered as broadcast packets. The slave's transmission of an ACL packet is controlled by a POLL packet sent from the master in the master-to-slave slot. When receiving the POLL packet, the slave responds by sending data packets in case it has packets to send, or in case it has nothing to send, it sends a NULL packet in the following slave-to-master slot. The bandwidth allocation for a slave's ACL link is controlled by the polling frequency of the master. The more frequently a slave receives poll packets, the more bandwidth a slave is allocated. Bluetooth specifications define various types of asynchronous data packets for the ACL link (i.e., DM1, DH1, DM3, DH3, DM5, DH5, AUX) which have different payload size, error detection scheme, error correcting scheme, and achievable data rate. The ACL link is used by asynchronous data packets and control packets (i.e., from LC, LM, L2ACP).

8.4.1.2   Logical Channels

Bluetooth defines five logical channels used to transfer different types of information:

- *Link Control (LC) Channel:* This channel carries low-level link control information such as flow control and payload characterization.
- *Link Manager (LM) Channel:* Typically it carries control information exchanged between the master's LMs and the slaves.
- *User Asynchronous (UA) Channel:* It carries L2CAP transparent asynchronous user data. These data may be transmitted in one or more baseband packets.
- *User Isochronous (UI) Channel:* It carries L2CAP transparent user isochronous data. These data may be transmitted in one or more baseband packets.
- *User Synchronous (US) Channel:* It carries transparent synchronous user data. This channel is carried over the SCO link.

LC and LM are used at the link control level and LM. UA, UI, and US are used to carry user information. The information on the LC channel is carried in the packet header, while the information for all other channels is carried in the packet payload. The information for the US channel is carried only by the SCO link. The information for the UA and UI channels is normally carried by the ACL link; however, it can also be carried by the data portion of the combined data-voice packet on the SCO link. The information on the LM channel can be carried either by the SCO or the ACL link. Figure 8.8 shows the flow of each logical channel.

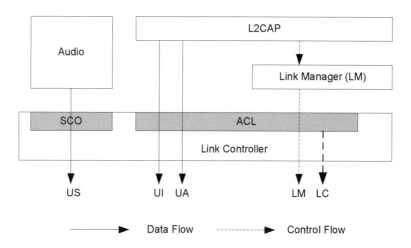

Figure 8.8 Logical Channels

8.4.1.3 Packet Format

Figure 8.9 shows the standard packet format, which includes the Access Code, LC header, and payload.

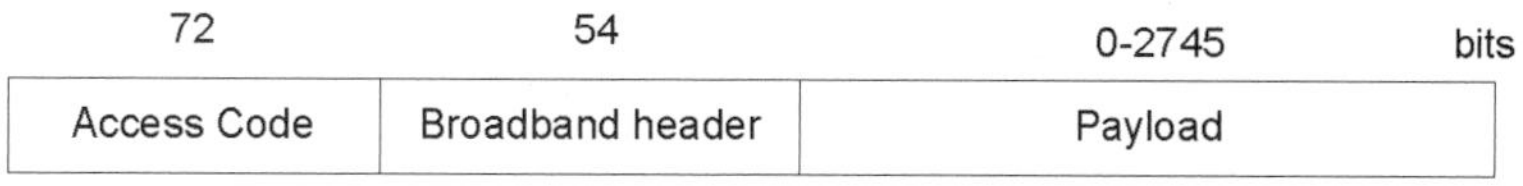

72	54	0-2745	bits
Access Code	Broadband header	Payload	

Figure 8.9 Standard Packet Format

There are three types of access codes:

- *Channel Access Code (CAC):* CAC is used to identify a piconet. CAC is determined by the device address of the piconet's master. Therefore, CAC is unique in each piconet. CAC is normally included in the user data packet.
- *Device Access Code (DAC):* DAC is used to identify a device. DAC is uniquely determined by the device address. DAC is used for the paging procedures. For example, in the paging procedure, the master will send the DAC of the slave with which it wants to establish a connection. Then, the slave responds back with its DAC. The DAC is included in the paging and paging response messages which include only the access code field (no header and payload).
- *Inquiry Access Code (IAC):* IAC is used for the inquiry procedures. IAC is included in the inquiry message which includes only the access code field (no header and payload). The master sends inquiry messages to collect the device addresses of slaves in the transmission range.

Figure 8.10 shows the packet format of the ACL and SCO links. An SCO packet contains the voice payload while an ACL packet contains the data payload. Some SCO packets can also contain data payload. The LC can use either SCO or ACL packet with or without payload (i.e., NULL and POLL packets do not have data payload while frequency hop synchronization (FHS) packets include control information in the data payload). The LM includes its control information in the data payload.

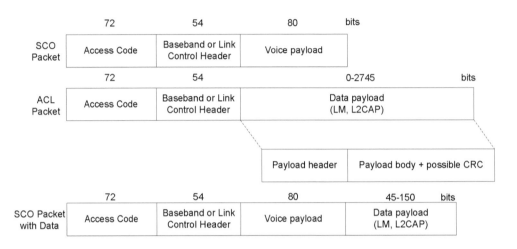

Figure 8.10 SCO and ACL Packet Format

8.4.2   Link Control

Before the master and slaves can communicate with each other, they have to establish the piconet using link control mechanisms based on link control states. Link control information is included in the link control header of the packet as shown in Figure 8.10.

8.4.2.1   Link Control States

A station can be in one of a number of link control states (see Figure 8.11). There are two major states: Standby and Connection. In addition, there are substates with their associated procedures: page, page scan, inquiry, and inquiry scan. These procedures are used to establish a piconet or add new slaves to a piconet. The substates are transient states between the Standby and Connection states.

8.4.2.1.1 Standby State   The station default state is Standby state, which is a low-power consumption mode. The Standby stations are not associated with any piconets. Occasionally, the Link Controller may leave this state to scan for page or inquiry messages, or to page or inquiry itself. If the station received such a message (i.e., page message, inquiry message), it will enter the Connection state and become a slave. If the station itself transmits a successful page message or inquiry message, then it will enter the Connection state and become the master.

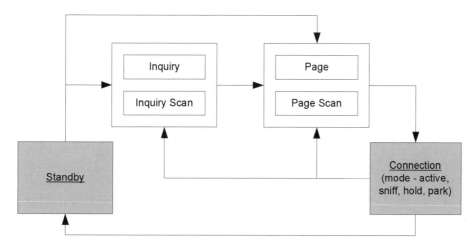

Figure 8.11 Link Control State Diagram

Inquiry procedures (see Figure 8.12) are the interaction processes between the station in the inquiry substate (called inquiry station) and the station in the inquiry scan substate (inquiry-scan station). Inquiry is used where the destination's device address is unknown to the source. The inquiry stations are the stations that want to discover which new stations are within their transmission range. The inquiry-scan stations are the stations that want to be discovered by other stations. The inquiry station transmits the inquiry message (containing IAC) with a frequency

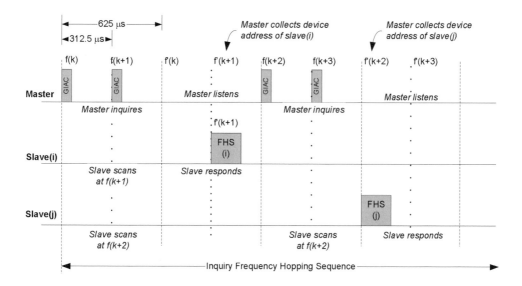

Figure 8.12 Inquiry Sequences

that follows the inquiry frequency hopping sequence. The inquiry-scan station scans the inquiry message at a single hop frequency. The inquiry-scan stations can receive the inquiry message whenever the inquiry station transmits at the same frequency. After receiving the inquiry message successfully, the inquiry-scan station will send to the inquiry station the response message that contains the device address and the clock. In summary, during this process, the inquiry station collects the device addresses and clocks of all stations that respond to the inquiry message. Based on this information the inquiry station can issue a page message to establish connection to known stations.

Paging procedures are used to establish new connections between the master and a slave. The paging substate is deployed by the master to establish the new connection with the slave which is in the paging-scan substate. The paging procedures are similar to the inquiry procedures. The master tries to connect to the targeted slave by repeatedly transmitting the slave's device access code (DAC) in several hop channels. Since the clocks of the master and the slave are not yet synchronized, the master does not know exactly when the slave wakes up and on which hop channel. The master predicts the hop channels where the slave should be, for example, by recalling details of the last exchange of packets. Based on this prediction, it transmits a series of the same DACs at the predicted channel as well as other channels and listens between transmissions for a response from the slave. In Page Scan, a slave listens for its own DAC for the duration of the scan window. During this window, the slave listens at a single hop frequency, its correlator matched to its device access code. When a slave enters Page Scan, it selects the scan frequency according to the page hopping sequence corresponding to this station's device address. Figure 8.13 depicts the paging sequences between the master and a slave. The master transmits paging messages more frequently (3200 packets per second or once every 312.5 μs) than the normal data packet transmissions (1600 packets per second or once every 625 μs). The master determines its paging frequency hopping sequence based on its device address. The slaves

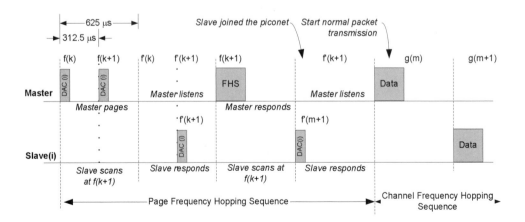

Figure 8.13 Paging Sequences

listen or scan at a fixed frequency. If a slave receives the paging message, it will transmit a paging response message which contains its DAC. Then, the master transmits the FHS message in order for the slave to synchronize 1) to the channel frequency hopping sequence and 2) to the clock. Then, the slave replies with a FHS packet and becomes an active node of the piconet.

In Standby, no connection has been established and the station can use all of its capacity to support Page Scan. If desired, the station can place ACL connections in the Hold mode or even use the Park mode before entering Page Scan. SCO connections are preferably not interrupted by Page Scan. In this case, Page Scan may be interrupted by the reserved SCO slots which have higher priority than Page Scan.

8.4.2.1.2 Connection State   At the connection state, a Bluetooth station becomes the member of a piconet. The station can be in one of the following four connection modes: Active, Sniff, Hold, and Park. These connection modes are used to save power and to allow stations to communicate with different piconets.

Active Mode: In this mode the station can actively communicate on the wireless channel. A piconet is allowed to have up to eight active mode stations at the same time. The master schedules the transmission based on traffic requirements to and from the slaves. In addition, the master supports regular transmissions to keep slaves synchronized to the channel. Active slaves listen during the master-to-slave slots. If an active slave is not addressed, it may sleep until the next new master transmission. A periodic master transmission is required to keep the slaves synchronized to the channel. Since the slaves only need the channel access code for synchronization, any packet type can be used for this purpose.

Sniff Mode: Slave stations can change to a power-saving mode in which the station's activity is decreased. In this Sniff mode the slave station listens to the piconet less often, thus reducing power consumption. If a slave participates on an ACL link, it has to listen to the master traffic every ACL slot. The time interval between consecutive listening events is programmable by the application.

Hold Mode: Slave stations can change into another power-saving mode. During the Connection state, the ACL link to a slave can be in Hold mode. This means that the slave temporarily does not support ACL packets while continuing to support possible SCO links. While in Hold mode, the station can execute procedures such as scan, page, inquire, or communicate with another piconet. Before the slave enters Hold mode, the master and slave have to agree on the time duration that the slave can remain in this mode. Once this time expires, the station will return to Active mode, synchronize to the traffic, and wait for instructions from the master.

Park Mode: In this mode the slave station is still synchronized to the piconet but does not communicate. Parked stations give up their active member address and only occasionally listen to the traffic from the master to synchronize and check for broadcast messages. To support parked slaves, the master establishes a beacon channel where one or more slaves are parked. The master transmits on this channel at constant time intervals a series of beacon slots.

8.4.2.2   Piconet Establishment and Operation

The piconet is governed entirely by its master. The master's address determines the frequency hopping sequence and the channel access code (the code that uniquely identifies the piconet). Since Bluetooth is based on TDD, every Bluetooth station has an internal system clock which determines the timing and hopping of the radio transceiver. This clock has no relation to the time of day and can therefore be initialized at any value, does not require adjustments, and is never turned off. For synchronization with other units, only offsets are used and added to the native internal clock. To ensure synchronization, the master sets the timing of all slaves based on its clock by transmitting to the slaves its clock reading. The slaves add an offset value to their native clocks so they can be synchronized to the master clock. Since the clocks are free-running, the offsets have to be updated regularly. Establishing piconet is controlled by link control state. Figure 8.14 shows the piconet establishment sequence.

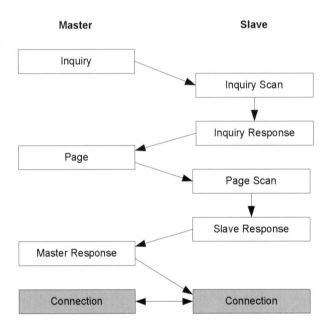

Figure 8.14 Piconet Establishment Sequences

First, the master executes the inquiry process to obtain the device addresses and clocks of the slaves within the transmission coverage. Then, the master performs the paging process to establish the connections with the specific slave. Finally, both the master and the active slaves can communicate with each other.

8.4.2.3   Scatternet Establishment and Operation

To increase the coverage area and the number of supported stations, several piconets may coexist in the same area. Each piconet's master specifies a different hopping series. In addition, the packets carried on the channels are preceded by different channel access codes as determined by the master addresses. As the number of coexisting piconets increases, the probability of collisions increases since different piconets may transmit on one of the 79 frequency channels. If several piconets coexist in the same area, a station can participate in two or more piconets by applying time multiplexing. In each piconet, the station will use this piconet hopping series, master device address, and proper clock offset. A Bluetooth station can act as a slave in several piconets but as a master in only one piconet. A group of connected piconets is referred to as a scatternet.

Stations must use time multiplexing to switch between the piconets in which they participate. In case of ACL links, a station can request to enter the Hold or Park mode so that during this mode it can join the other piconet. Units in the Sniff mode may have sufficient time to visit another piconet between the sniff slots. If SCO links are established, other piconets can only be visited in the non-reserved slots between the sniff slots. Since coexisting piconets do not have a mechanism to synchronize clocks, additional guard time should be considered for proper switching.

A master can become a slave in another piconet by being paged by the master of this other piconet. On the other hand, a slave participating in one piconet can be a master in another piconet. Since the paging unit always starts out as master, a master-slave role exchange will be required. Such an exchange may involve a reorganization of piconets since slaves may change from the current master to the new master. This is a complex process since it requires all stations to resynchronize their clocks, transit to the frequency hopping series of the new masters, and adjust to the new master device addresses. This reorganization of piconets should be done without interfering with ongoing communication activities and without involving the users. The option of setting up a completely new set of piconets will require a long period of time. Another option is to have the new masters utilize their knowledge of timing and hopping sequences to reduce the time for such an exchange. Because of the complexity of the piconet reorganization process, it is currently left outside the scope of the Bluetooth standard. Therefore, Bluetooth developers need to design and implement their own exchange algorithm.

8.4.2.4   Link Supervision

Several events can cause stations to loose their communication link. Such events can be power failure, movement out of range, and interference from outside sources. Hence, the standard defines a process in which the link is monitored by the master and the slave using link supervision timers for both SCO and ACL connections. When a station receives a packet with a valid slave address, this timer is reset. However, when this timer reaches a predetermined value when the station is in the Connection state, the connection is reset. This timer timeout period is negotiated at the Link Management (LM) such that it is longer than Hold and Sniff periods. If the slave is in Park mode, then link supervision is done by un-parking and re-parking the slave.

8.5 Link Manager (LM)

The Link Manager is responsible for link setup, security, authentication, link configuration, timer setup, and other control procedures (Figure 8.15). The communication between the stations' LM modules is done using the Link Manager Protocol (LMP). The LM sends its control message through Data Medium rate (DM1) or Data Voice (DV) packets. Due to the fact that DM1 and DV packets use reliable links (using 2/3 FEC and CRC), there are no explicit acknowledgment messages. LM messages have higher priority than user data.

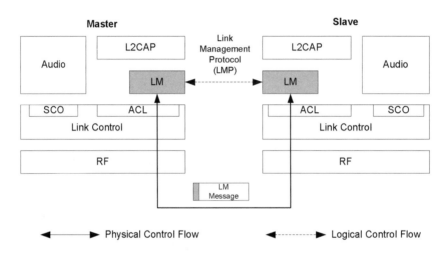

Figure 8.15 Link Manager

LM supports upper layer applications that execute control algorithms such as managing the connection states (Park, Hold, Sniff, Active), establishing security, and supporting QoS. These control algorithms may be self contained or may allow input from users. We will discuss more about the QoS aspects of the LM protocols in a later section.

8.6 Host Control Interface (HCI)

As we showed in Figure 8.4, which describes the Bluetooth protocol stack, HCI is the layer between L2CAP and LM. The HCI provides a standard interface to Bluetooth so that upper layers can be independent and transparent from the host hardware implementation. It provides interface to the LM and access to the hardware status and control registers. HCI consists of multiple parts (i.e., HCI driver, HCI firmware) that reside in both Bluetooth Host and Bluetooth hardware. Figure 8.16 shows HCI relationship to the Bluetooth system.

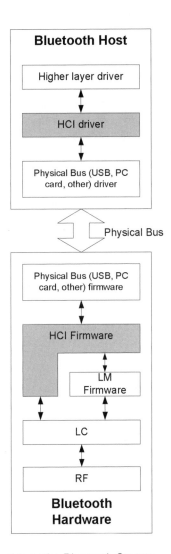

Figure 8.16 HCI Relationship to the Bluetooth System

8.7 Logical Link Control and Adaptation Protocol (L2CAP)

The Logical Link Control and Adaptation Layer Protocol (L2CAP) provides connection-oriented and connectionless data services to upper layer protocols. The L2CAP specification is defined for only Asynchronous Connectionless (ACL) links. Support for Synchronous Connection-Oriented (SCO) links, mainly used for real-time voice traffic with reserved bandwidth, is not yet provided in L2CAP. Bluetooth recommends that L2CAP should include simple and low-overhead programs so they can fit in devices with limited computational, power, and memory resources.

These data services include QoS support, group support, and protocol multiplexing, segmentation and reassembly. Protocol multiplexing enables an application to simultaneously use several higher layer protocols such as TCP/IP and RFCOMM.

L2CAP labels each connection by a channel identifier (CID) and assumes that each channel is a full duplex connection and may have a QoS flow specification. The L2CAP connection establishment process allows exchange of the expected QoS between two Bluetooth stations. Moreover, each station monitors the resources used to ensure that the required QoS contract is provided. Again, these QoS algorithms are left to the developer and are not provided as part of Bluetooth.

L2CAP group support permits higher level applications to map groups (group of addresses) to piconets. The Baseband layer supports the concept of a piconet for a unique group. Without group support, higher level applications would need to establish a complex communication process directly with the Baseband Protocol and LM layers.

8.8 Higher Bluetooth Layers

8.8.1 RFCOMM

RFCOMM enables compatibility with applications that use the serial port as their main communication bus. RFCOMM conveys all of the RS232 control signals and supports remote port configuration.

8.8.2 Service Discovery Protocol (SDP)

The service discovery protocol (SDP) provides tools for applications to discover which services are available and to determine the nature of these services.

8.8.3 Audio and Telephony Control (Tel Ctrl)

This layer includes the interface needed to connect and disconnect a telephone call, including signaling the devices that participate in the connection. Telephony audio links are established with synchronous links and therefore do not go through the same L2CAP-to-LM path that asynchronous links go through. In other words, audio links may be thought of as direct Baseband-to-Baseband links.

8.9 Profiles

Bluetooth profiles specify a set of basic standards to ensure device interoperability. In Bluetooth Specifications, v. 1.1, there are 13 Bluetooth profiles. They include profiles such as Generic Access, Cordless Telephony, Intercom, Serial Port, Headset, Dial-up Networking, Fax, and LAN Access and Synchronization (exchange of information between applications such as calendar).

8.10 QoS Support

Bluetooth aims to support both synchronous and asynchronous data services. The key concept introduced by Bluetooth for QoS support is its definition of two data link types, SCO and ACL. Furthermore, Bluetooth also defines QoS as enabling MAC and QoS signaling. In this section, we take a closer look into Bluetooth's QoS mechanisms. Figure 8.17 shows Bluetooth's QoS architecture.

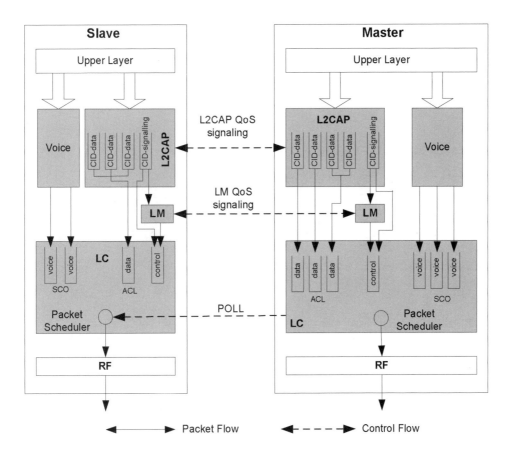

Figure 8.17 Bluetooth QoS Architecture

8.10.1 Classification

Bluetooth defines packet classification at the L2CAP layer. All asynchronous data packets are assigned a CID as shown in Table 8.1.

Table 8.1   Channel Identifiers

CID	Description
0x0000	Null identifier
0x0001	Signaling channel
0x0002	Connectionless reception channel
0x0003 – 0x003F	Reserved
0x0040 – 0xFFFF	Dynamically allocated

For a connection-oriented application, CIDs are dynamically assigned on both end stations. For a connectionless application, a fixed CID (0x0002) is assigned to the receiver and a dynamic CID is allocated to the sender. For L2CAP signaling, both end stations are allocated CID=0x0001. A classifier will use the CID to identify the packets and forward them to the appropriate queue. Bluetooth does not define how to classify the voice packets that do not go through the L2CAP layer.

8.10.2 Channel Access

As described in Section 8.4, Bluetooth deploys a TDD channel access. The master is the only station that transmits on the downlink direction (master-to-slave time slot). On the other hand, the uplink (slave-to-master time slot) is shared by multiple slave stations. The uplink channel access is controlled by the master which allocates reserved time slots for SCO packets and uses POLL control packets for ACO packets. After receiving a POLL, the slave is allowed to transmit ACO packets. Bluetooth uses a collision-free channel access scheme which provides tight channel access control. Therefore, these schemes can provide QoS support for applications with strict QoS requirements.

8.10.3 Packet Scheduling

We differentiate between two packet scheduling schemes: Intrastation packet scheduling and Interstation packet scheduling. There are three types of packets: SCO packets, ACL packets, and Control packets (which use the ACL link). In the LC module of the slave, a slave can have up to three SCO links for the same master or up to two SCO links for different masters, one ACL link and one control link. In the LC module of the master, the master can have up to three SCO links, one ACL link for each slave, and one control link. Using these packet scheduling mechanisms, Bluetooth can provide strict QoS service for voice applications using the SCO link and loose QoS service for asynchronous applications using the ACL link.

8.10.3.1   Intrastation Packet Scheduling

SCO packets are transmitted in the reserved time slot and ACL packets are transmitted into the rest of the time slots. Therefore, Bluetooth implicitly deploys strict priority packet scheduling. SCO has higher priority than ACL. Furthermore, Bluetooth also explicitly defines that control packets have higher priority than ACL packets. In summary, SCO has the highest priority followed by control packets and ACL packets. There is only one ACL link in a slave. Asynchronous data packets from different applications initiated by the same slave will be aggregated in the same queue. All packets experience the same QoS. Moreover, it is possible that an asynchronous application can abuse the other asynchronous applications by using the bandwidth as much as it can.

8.10.3.2   Interstation Packet Scheduling

The slave is allowed to transmit in the reserved time slot (for SCO) or in response to a POLL message (for ACL). The bandwidth allocation for SCO packets depends on the number of reserved time slots which are assigned by the signaling process. The bandwidth allocation for ACL packets depends on the number of POLL packets received (polling frequency) which is controlled by the master. Bluetooth does not define the polling algorithm which determines the polling sequence and polling frequency.

8.10.4 QoS Signaling

There are two levels of QoS signaling: L2CAP QoS signaling and LM QoS signaling.

8.10.4.1   L2CAP QoS Signaling

L2CAP QoS signaling exchanges the QoS information between L2CAP of master and slave. Such QoS information indicates the traffic characteristic and QoS requirements of applications (called FlowSpec) such as Token Rate, Token Bucket Size, Peak Bandwidth, and Latency and Delay variation. The QoS information is included in the Configuration Parameter Option field of the L2CAP control message. L2CA_ConfigReq is the L2CAP control message that performs the connection request. L2CA_ConfigReq requests the traffic characteristics and QoS requirement of outgoing data flow to the master. Then the master sends L2CA_ConfigRsp in response to L2CA_ConfigReq. L2CA_ConfigRsp indicates the accepted traffic characteristics of the incoming data flow. Bluetooth does not define the admission control process that decides the accepted traffic characteristics. Bluetooth does not define the algorithm that maps the FlowSpec to polling frequency and polling sequence. Another L2CAP QoS signaling message is L2CA_QoSViolationInd, which is used for indicating the address of the remote device that violates a QoS contract.

8.10.4.2   LM QoS signaling

Bluetooth also defines QoS signaling in the LM layer. For example, LMP_quality_of_service and LMP_quality_of_service_req contain the polling interval for the ACL link. These two messages are used for the request and response of the polling interval of the ACL link.

CHAPTER 9

IEEE 802.15

The IEEE 802.15 group focuses on standards that will cover Wireless Personal Area Networks (WPANs). So far IEEE 802.15 has introduced one standard, referred to as IEEE 802.15.1, which standardized parts of Bluetooth. In addition, the group works on additional standards that will include a high data rate WPAN, referred to as IEEE 802.15.3, and a low data rate WPAN, referred to as IEEE 802.15.4. The group is also developing recommended practices to facilitate the coexistence of IEEE 802.15 and IEEE 802.11, referred to as IEEE 802.15.2.

9.1 IEEE 802.15.1

IEEE 802.15.1 standard adopted the Bluetooth Medium Access Control (MAC) and Physical Layer (PHY) specifications in early 2002. It adopted Bluetooth Version 1.1 specifications for lower transport layers (L2CAP, Link Manager Protocol, and Baseband) and radio (see Figure 9.1).

IEEE 802.15.1 PHY and MAC key mechanisms are almost identical to the mechanisms introduced in the Bluetooth specifications. IEEE 802.15.1 specifies the Logical Link Control (LLC) which supports the LLC function. Therefore, the readers can refer to appropriate sections of Chapter 8 for information regarding the physical layer, Baseband, link manager, Logical Link Control and Adaptation Protocol (L2CAP), and Host Control Interface (HCI) in IEEE 802.15.1.

9.2 IEEE 802.15.3

The IEEE 802.15 group is working on a high data rate WPAN that can support connectivity needs between portable devices. This section provides details on a potential future standard. However, the future standard may be completely different from this description.

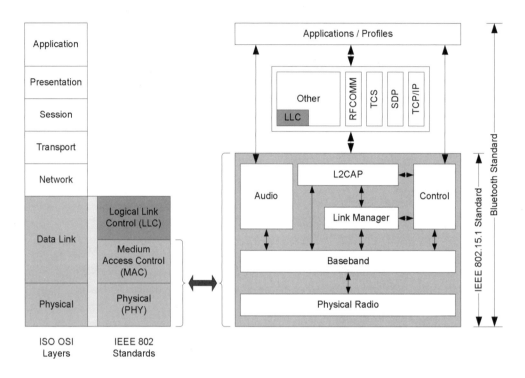

Figure 9.1   Mapping of ISO OSI to IEEE 802.15.1 WPAN Standard

The working group is motivated by the need to distribute video and audio that requires much more bandwidth than IEEEE 802.15.1 can sustain. The proposed radio is based on single carrier Quadrature Amplitude Modulation (QAM) with Trellis Coded Modulation (TCM) working in the 2.4 GHz band. Several selectable speeds are discussed such as 11, 22, 33, 44, and 55 Mbps along with three to four non-overlapping channels. IEEE is also considering another radio technology based on Ultra Wide Band (UWB) that will allow speeds of around 100 Mbps within a 10 meter range and 400 Mbps within a 5 meter range. In addition to data applications, they envision applications such as: high-speed transfer of digital video from a digital camcorder to a TV or PC, interactive games that are media rich, home theater, and delivery of multimedia from a PC to an LCD projector. The MAC provides support for multimedia QoS requirements via a TDMA based superframe architecture with Guaranteed Time Slots (GTSs).

9.2.1   IEEE 802.15.3 Architecture

9.2.1.1   Network Topology

The architecture is based on the concept of a piconet which is established ad hoc and allows a number of independent data devices (DEVs) to establish peer-to-peer communication (see Figure 9.2). It allows mobile devices to join and leave the piconet with short association times. The piconet is confined to a personal area and typically is expected to cover a range of at least 10 meters and possibly up to 70 meters. A piconet is established once a DEV that is capable of becoming the Piconet Coordinator (PNC) begins to transmit control messages called beacons. Through the beacon, the PNC maintains network synchronization by providing timing information. The PNC is also responsible for controlling the channel access, managing the QoS requirements, performing admission control, and assigning time slots for connections between DEVs. Before transmission of data in a piconet, a DEV is required to associate with the PNC and follow the channel access information provided in the beacons.

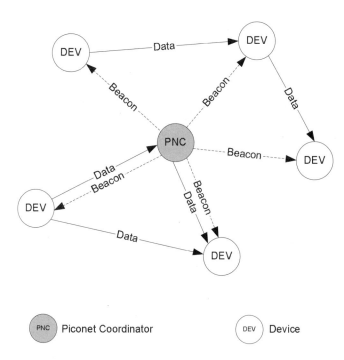

Figure 9.2   IEEE 802.15.3 Piconet Elements

Several piconets can coexist by sharing the same channel frequency. The piconets are combined to cover a larger area and include more DEVs (see Figure 9.3). There is a proposal for a child piconet to be established for an existing piconet. Hence, the child piconet can extend the area of coverage of the existing piconet, increase the number of DEVs to be supported, or shift some tasks to the new piconet. When such a child piconet is established, the existing piconet becomes the parent piconet. The child piconet uses a unique piconet ID (PNID) and functions almost independently. The child PNC handles association, authentication, security, and acknowledgments without the parent PNC. The DEVs that associate to the child piconet can communicate with each other within the child piconet. However, to avoid the interference between the parent piconet and child piconets, the parent PNC has to allocate GTSs or reserve the time slot for the child piconet to access the channel. DEVs of the child piconet are allowed to exchange data only within the allocated GTS. The child piconet's PNC is also a member of the parent piconet and thus can communicate with the DEVs in both the parent piconet and its own child piconet.

The committee also proposed the idea of establishing a neighbor piconet under an existing piconet that will become the parent piconet. This will provide the mechanism for sharing the frequency spectrum between different piconets when there are no vacant physical radio channels. The neighbor piconet will use a unique PNID. Similar to the child piconet, it will function almost independently from the parent piconet. It will handle association, authentication, security, and acknowledgments without involving the parent piconet. However, the parent piconet has to allocate GTSs for the neighbor piconet to access the channel. Unlike the PNC of the child piconet, the neighbor's PNC is not a member of the parent piconet. The neighbor's PNC is allowed to send to the parent piconet only certain commands such as association request, disassociation, channel time request (CTR), and authentication. The creation process of child and neighbor piconets is described in Section 9.4.2.

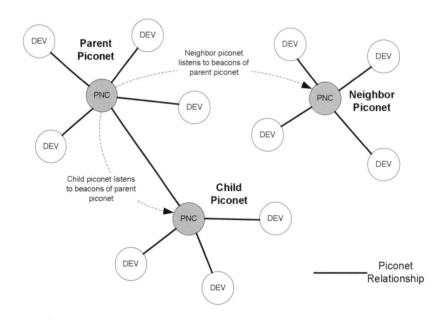

Figure 9.3 Parent/Child/Neighbor Piconets

9.2.1.2   Protocol Stack

As shown in Figure 9.4, the draft standard divided the protocol stack into various layers: frame convergence sublayer (FCSL), MAC sublayer, PHY sublayer and layer-dependent management entity (MLME, PLME). The frame convergence sublayer provides frame multiplexing from different protocols (i.e., 802.2, 1394, USB, etc.). The MAC sublayer defines the following functionalities: association/disassociation, channel access, channel time management, channel synchronization, fragmentation/defragmentation, acknowledgment/retransmission, peer discovery, multirate support, dynamic channel selection, and power management. The PHY sublayer defines the PHY specifications. The Device Management Entity (DME) which is a layer-independent entity provides the status of various layer management entities. The standard does not detail the DME functions, but rather provides a general overview of its functionality. Each entity interacts through a Service Access Point (SAP). Some SAPs (i.e., MAC SAP, PHY SAP, MLME SAP, MLME-PLME SAP) are explicitly defined within the draft standard but some, such as the interface between the MAC and the MLME as well as the interface between PHY and PLME, are not explicitly defined.

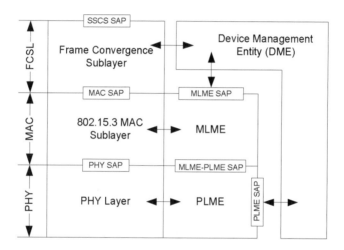

Figure 9.4   IEEE 802.15.3 Protocol Stack

9.3 IEEE 802.15.3 Physical Layer

IEEE 802.15.3 operates in the 2.4 to 2.4835 GHz frequency band. As shown in Table 9.1, there are a total of five channels in two sets of channel assignments: high-density set and 802.11b coexistence set. Each channel has 15 MHz bandwidth. For the high-density set (802.11b does not coexist) four channels are defined while three channels are defined in 802.11b coexistence set. If a DEV detects 802.11b network operating in its area, it should use 802.11b coexistence channel set.

The PNC has the ability to select a channel for the piconet operation. Using information such as interference from other users, other IEEE 802.15.3 piconets, and other unlicensed wireless entities (i.e., IEEE 802.11b) it can make a decision on the best channel for its piconet. The PNC has the capability to dynamically and transparently change the channel on which the piconet is operating without requiring either user intervention or interruption of ongoing applications.

Table 9.1   2.4 GHz Channel Assignment

Channel ID	Center frequency	High-density channel set (802.11b does not coexist)	802.11b coexistence channel set
1	2.412 GHz	X	X
2	2.428 GHz	X	
3	2.437 GHz		X
4	2.445 GHz	X	
5	2.461 GHz	X	X

The DEVs can discover information about the services and capabilities of other DEVs in the piconet. This information can be collected in several ways: the PNC information request command, the probe command, and the piconet services information element within the beacon. In addition, the PNC can ask a DEV in its piconet to evaluate the wireless media conditions either in the current channel or in an alternate channel using the remote scan request command. A DEV can ask another DEV in the piconet about the status of the current channel using the channel status request command. The PNC information request command is used to obtain information from the PNC about either a specific DEV or all of the DEVs in the piconet. The PNC reply includes the ID and address of the DEV, its supported data rates, an indication whether this DEV is PNC capable, an indication whether it is a PNC of another piconet, and other information.

The PHY supports various data rates corresponding to the modulation and coding used such as 11 Mbps (QPSK[Quadrature Phase Shift Keying]-TCM), 22 Mbps (QPSK), and 33, 44, 55 (16-32-64 QAM-TCM, respectively).

Recently, IEEE 802.15.3 established the 802.15.3a study group with the charter to investigate alternate PHY to support very high data rate applications (more than 100 Mbps). Examples of such applications are: 1394a/USB 2.0 high-speed cable replacement, high-density DVD, high-resolution printer and scanner, and digital camera. Ultra Wide Band (UWB) is a potential PHY candidate for such very high data rate applications.

9.4 IEEE 802.15.3 Media Access Control

IEEE 802.15.3 MAC includes several functions. In this subsection we will describe only some key mechanisms such the channel access and the creation of a piconet.

9.4.1 Channel Definition and Channel Access

The MAC is based on a time-slotted superframe structure that consists of three periods: Beacon, Contention Access Period (CAP), and Contention-Free Period (CFP) (see Figure 9.5). Within a CFP, there are two types of time slots: *management time slot (MTS)* reserved for command exchange between PNC and DEVs, and *guaranteed time slot (GTS)* reserved for data exchange among DEVs. The length of each period or time slot is determined by the PNC via beacons. At the beginning of each superframe, the PNC broadcasts beacons to specify the control, resource allocation, and time synchronization to the entire piconet.

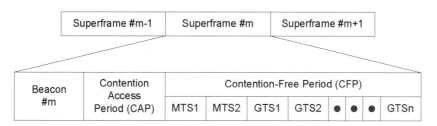

Figure 9.5 Superframe Structure

9.4.1.1 Contention Access Period (CAP)

In the CAP, contention is allowed via the carrier sense multiple access with collision avoidance (CSMA/CA) mechanism. The DEVs can send small amounts of pending data without requesting reserved time slots. Because of the unpredictable channel access delay of the contention scheme, CAP is suitable for asynchronous data traffic such as file transfer, MP3 download files, etc. Moreover, CAP can be used for command exchange between the PNC and the DEVs. CAP starts after the end of beacon and terminates at a time specified by piconet synchronization parameters information element in a beacon. Figure 9.6 shows the CAP within the superframe.

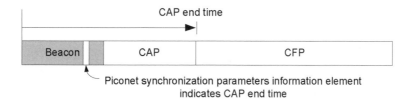

Figure 9.6 CAP Structure in the Superframe

The draft standard defines two interframe spacings (IFSs) used in CSMA/CA mechanism: *short interframe space (SIFS)* and *retransmission interframe space (RIFS)* with RIFS = SIFS + aBackoffSlot. Both SIFS and aBackoffSlot duration are PHY dependent. The CSMA/CA mechanism

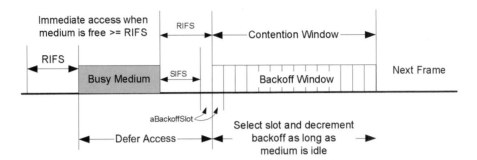

Figure 9.7 CSMA/CA in CAP

and backoff procedure are similar to the IEEE 802.11 MAC introduced in Chapter 4 except for the different IFSs (i.e., RIFS is analogous to DIFS in IEEE 802.11). Figure 9.7 shows the CSMA/CA mechanism in CAP.

A DEV transmits a data packet after it senses the channel idle for a time duration equal to RIFS. Acknowledgment packets are transmitted immediately after SIFS time. When a collision occurs (i.e., multiple DEVs transmit packets at the same time), the exponential backoff algorithm (i.e., doubling the contention window) is applied.

9.4.1.2   Contention-Free Period (CFP)

The CFP is based on time division multiple access (TDMA) in which the PNC guarantees the starting time and reserved duration of time slots for traffic streams that require certain amounts of bandwidth. Time slot allocation is specified by the *channel time allocation (CTA)* information element (IE) in a beacon. As shown in Figure 9.8, the CTA information element includes source and destination addresses of the DEVs that transmit and receive the traffic stream, the stream index that identifies the traffic stream, and the starting time and duration of the reserved time slots (MTS or GTS).

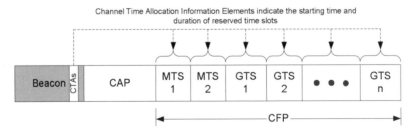

Figure 9.8 Channel Time Allocation (CTA)

Since the bandwidth is allocated using a TDMA scheme, the channel is fully controlled by the PNC and can provide QoS support. Therefore, CFP is suitable for isochronous data traffic such as video streaming, audio streaming, and video/audio broadcast. We will discuss QoS aspects in

Section 9.5. As mentioned earlier, CFP includes two types of time slots: MTS and GTS. MTS is used for command exchange while GTS is used for data exchange. GTS can accommodate not only isochronous data traffic but also asynchronous data traffic. However, the ways to request and allocate time slots are slightly different between the two types of slots.

The PNC allocates time slots through the CTA mechanism. A source DEV, which has pending data traffic and is seeking reserved time slots for transmitting the data, first sends the *channel time request (CTR) command* to the PNC to indicate the recurring duration and the number of required time slots. For isochronous data traffic, the CTR command contains 1) the number of time slots needed (i.e., the number of time slots per superframe) and 2) the duration of each time slot (i.e., the minimum duration and desired duration). If there are available resources or time slots to accommodate the new request, the PNC allocates GTSs to the requesting DEV by including them in the CTA information element. On the other hand, for asynchronous data traffic, the CTR command contains the total amount of time needed to transmit the data instead of the recurring channel time. The PNC will schedule such time based on the channel availability. The source DEV, destination DEV, and PNC are all allowed to terminate the data connection whenever they want for both isochronous and asynchronous data traffic.

Each GTS is actually one time slot reserved for a specific DEV. This DEV may or may not make use of all the allocated time slots. Each DEV is responsible for determining how to use its reserved slot—that is, it can determine what command, stream, and asynchronous data will be transmitted.

There are two types of GTSs: dynamic GTS and pseudostatic GTS (see Figure 9.9). The type of GTS is also indicated in the CTR command. The PNC may dynamically change the location of the dynamic GTSs within the superframe, on a superframe-by-superframe basis. This provides a tool for the PNC to reassign GTS locations to optimize channel utilization. Pseudostatic GTSs, which are allocated only for isochronous data traffic, have fixed location (referenced to the beginning of the beacon) within the CTA. The channel time location of the pseudostatic GTS can be changed (but less frequently than dynamic GTS) to optimize the channel utilization. To carry out this change, the PNC has to go through a series of processes to notify the involved DEVs (both transmitting and receiving DEVs). The PNC notifies the DEVs by sending a special command (probe command) with the new location. After all involved DEVs have been confirmed, the new CTA is included in the beacon.

Management Time Slots (MTSs) are identical to GTSs. MTSs are used for exchanging commands between the PNC and the DEVs. There are two ways to deploy MTSs: an MTS for the command service of an individual DEV or an MTS for command service of multiple DEVs. In the latter case, the MTS is denoted as open MTS in which any DEV can transmit the command messages. The PNC determines the number of MTSs per superframe. Since multiple DEVs can send command messages in the same MTS, collisions can occur. Therefore, in open

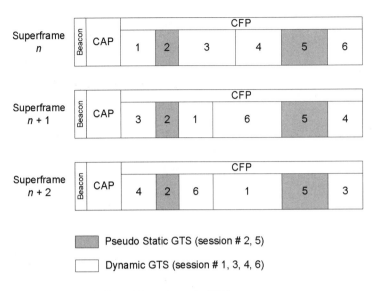

Figure 9.9 Dynamic GTS and Pseudostatic GTS

MTS the slotted Aloha protocol is used for channel access and contention resolution. Each DEV maintains its own contention window (CW) and a counter. The contention window is based on the number of retransmission attempts of a DEV. The contention window is defined as follows:

$$CW = \begin{cases} 256 & for\ 2^{a+1} \geq 256 \\ 2^{a+1} & for\ 2^{a+1} \leq 256 \end{cases}$$

$$where\ \ a = number\ of\ retransmission\ attempts$$

Each DEV assigns the counter value by randomly selecting an integer in the interval [1,CW]. The DEV counts down in each open MTS continuously across superframes. When the counter reaches 1, the DEV will transmit the command message in the current open MTS as shown in Figure 9.10. An acknowledgment (ACK) is required to indicate the success of the command transmission. An open MTS that is used for the association process is called association MTS.

9.4.2 Starting a Piconet

The creation of a piconet begins with a DEV that is capable of being a PNC. This DEV scans the channel and selects the channel that has the least interference and is not being used by other piconets. If the channel is available, the DEV becomes a PNC and starts sending beacons. Thus a piconet is established. If no channels are available, the DEV can try to establish a child or neighbor piconet instead.

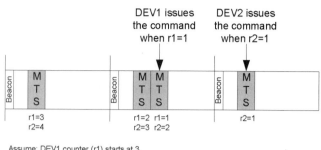

Assume: DEV1 counter (r1) starts at 3
DEV2 counter (r2) starts at 4

Figure 9.10 MTS Channel Access

9.4.2.1   Creating a Child Piconet

A PNC capable DEV establishes a child piconet to extend the coverage area of the existing piconet, which eventually becomes the parent piconet. The DEV sends the channel request command
to the parent's PNC to request the special GTS (called private GTS) for the child piconet. If a
time slot is available, the parent's PNC will assign a private GTS to the child piconet. The DEV
becomes the child's PNC and starts transmitting beacons. All communication of the child piconet takes place within the assigned private GTS as shown in Figure 9.11. Because the child's
PNC is also a member of the parent piconet, the parent's PNC may allocate GTS of the parent
piconet (which is also in unassigned GTS duration of the child piconet) to the child's PNC to
enable interpiconet communication between the child's PNC and any DEVs (including a PNC)
of the parent's piconet.

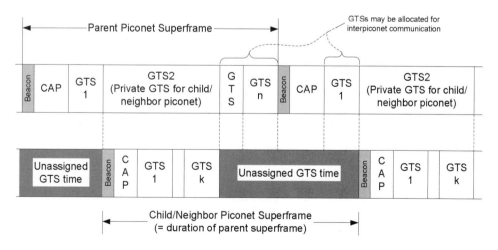

Figure 9.11 Parent Piconet and Child/Neighbor Piconet Superframe Relationship

9.4.2.2   Creating a Neighbor Piconet

If no channels are available during the piconet creation procedure, a PNC capable DEV can establish a neighbor piconet on the existing piconet, which eventually becomes the parent piconet. First the DEV sends an association request to the parent's PNC. If the neighbor association request is accepted, the DEV sends the channel request command to the parent's PNC to request private GTS for the neighbor piconet. Then the parent's PNC will assign a private GTS to the neighbor piconet. The DEV becomes the neighbor's PNC and starts transmitting beacons. The superframe relationship between the parent piconet and the neighbor piconet is the same as the relationship between the parent piconet and the child piconet as shown in Figure 9.11. Even though the neighbor's PNC is not a member of the parent piconet, the parent's PNC may allocate GTSs of the parent piconet (which is also in unassigned GTS duration of the neighbor piconet). This enables the neighbor's PNC to have limited communication between the neighbor's PNC and the parent's PNC such as association requests, disassociation requests, channel time request (CTR), and authentication requests.

9.5 IEEE 802.15.3 QoS Support

IEEE 802.15.3 aims to support both asynchronous and isochronous data services. The draft standard defines some QoS mechanisms to facilitate those services. IEEE 802.15.3 MAC focuses on channel access mechanisms. It leaves QoS negotiation and traffic handling to the upper protocol layers. As we know, isochronous data is time sensitive and requires certain bandwidth and delay commitment. As described in the previous section, IEEE 802.15.3 MAC has two modes of operation: Contention Access Period (CAP) and Contention-Free Period (CFP).

9.5.1   QoS in Contention Access Period

A DEV with pending data is not required to establish any connection or request any resource. A DEV accesses the channel and competes for bandwidth with other DEVs using the CSMA/CA mechanism. Therefore, CAP does not provide any service assurance—that is, it offers best effort service. CAP is suitable for asynchronous data services.

9.5.2   QoS in Contention-Free Period

Using TDMA, the PNC has full control over the channel by allowing a DEV to access the channel in a specific time for a specific duration. CFP provides certain QoS assurances in terms of bandwidth and delay. It aims to support high-bandwidth asynchronous data and isochronous data. A priori to transmission, a connection needs to be established between the DEV and the PNC. Such a connection is established by sending the PNC a channel time request (CTR) command. In case there are available resources, the PNC allocates GTSs. The connection establishment process is shown in Figure 9.12.

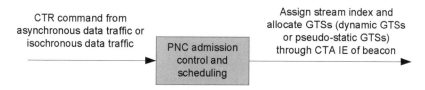

Figure 9.12 Connection Establishment Process

In establishing isochronous connections, the DEV must first determine the bandwidth and delay requirements of its isochronous connection in terms of the following parameters: the required number of time slots (GTSs) per superframe, the minimum and desired duration of each time slot, priority (based on eight priority values of IEEE 802.1p traffic types, see Table 9.2, and the desired GTS type (i.e., dynamic GTS or pseudostatic GTS). These parameters are included in the CTR command. The mapping process between the QoS requirements of an isochronous connection and the CTR parameters is the task of each DEV and it is not specified by the draft standard. The PNC admits and allocates GTSs based on the CTR information.

Table 9.2 IEEE 802.1p Priorities and Traffic Types

User priority	Traffic type	Used for	Comments
0 (default)	Best effort (BE)	Asynchronous data	Default piconet traffic
1	Background (BK)	Asynchronous data	
2	-	A spare	Currently not assigned
3	Excellent effort (EE)	Isochronous data	For valued customer
4	Controlled load (CL)	Isochronous data	
5	Video (VI)	Isochronous data	< 100 ms delay and jitter
6	Video (VO)	Isochronous data	< 10 ms delay and jitter
7	Network control (NC)		

In establishing asynchronous connections, the parameters included in the CTR command are slightly different from those in the isochronous connections. The DEV requests the total amount of time for the connection (i.e., total file transfer time) and the priority.

After receiving the CTR command, the PNC performs admission control to check if there are available resources. When there are available resources, the PNC allocates GTSs to the requesting DEV. The admission control and scheduling algorithms are not specified in the draft standard.

9.5.2.1   Classification

The draft standard specifies the stream index used to identify the connection or traffic stream (see Figure 9.13). The stream index is assigned by the PNC during the connection establishment procedure. There are two reserved stream indices: 0x00 for unassigned streams and 0xFE for all asynchronous streams. A stream index, except these two reserved stream indices, is uniquely assigned for each isochronous stream in the piconet. The classification process uses the stream index to identify the traffic and forward it to the appropriate queue. Due to the fact that there is only a single stream index for asynchronous streams, multiple asynchronous connections within a DEV will be aggregated to a single queue and wait for scheduling. Therefore, there is no service differentiation among asynchronous connections residing in the same DEV. On the other hand, each isochronous stream is assigned a unique stream index. This enables per-flow classification.

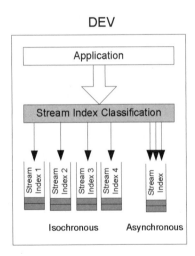

Figure 9.13 Classification

In summary, the TDMA channel access scheme combined with flow classification will enable a certain level of QoS support for isochronous traffic streams. However, there are a number of QoS mechanisms (i.e., admission control, packet scheduling) which are still undefined. The network designer may implement per-flow QoS solutions for which the standard defines some necessary QoS mechanisms such as per-flow (stream index) classification, CTA IE, and CTR command. The draft standard also provides optional priority which may be used to implement differential services. Figure 9.14 shows the simplified diagram of the IEEE 802.15.3 QoS architecture.

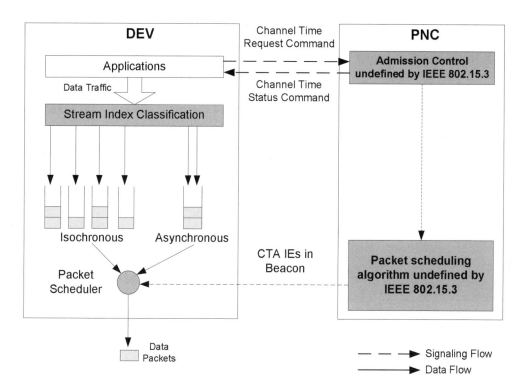

Figure 9.14 IEEE 802.15.3 QoS Architecture

9.6   IEEE 802.15.4

The IEEE 802.15.4 focuses on the development of a standard that details the wireless Medium Access Control (MAC) and Physical Layer (PHY) specifications for a Low-Rate Wireless Personal Area Network (LR-WPAN). The goal is to introduce a low-rate WPAN that is simple and low-cost that supports wireless network connectivity in applications that require minimal throughput as well as low power consumption. In this section we present some of the discussions and proposals in the group regarding the future standard. However, the future standard may be completely different.

The group may also address location tracking capabilities that can be applied within smart tags and badges. Other applications of the IEEE 802.15.4 standard may include personal networks, home networks, automotive networks, industrial networks, interactive toys, remote sensing, and cable replacement for the last meter connectivity. Sensor networks may include hundreds or thousands of wireless sensors. Examples of sensor networks include wearable personal health monitoring devices, home automation networks, and industrial control networks.

9.6.1   IEEE 802.15.4 Architecture

9.6.1.1   Network Topology

LR-WPAN consists of two or more interconnected devices (DEVs) that form a personal area network, which sometimes is called a personal operating space (POS). In the draft standard, the DEVs are categorized into two types: a Full Function Device (FFD) and a Reduced Function Device (RFD). The FFD can serve either as a network coordinator or as a simple network node. A network node (which typically runs the applications) operates as either a source or a destination of data traffic, whereas the network coordinator not only has the network node's functionality but also has the ability to route the data traffic, to control the channel access of other devices, and to provide basic timing within the POS. An FFD can communicate with other RFDs or FFDs and can control the network if it operates as a network coordinator. An RFD can only be a network node and can communicate only with a network coordinator (there is no direct connection between RFDs). The RFD is geared for devices that are extremely simple and do not need to send large amounts of data. Examples of such devices are light switches, actuators, smart badges, and passive infrared sensors. DEVs can be organized in three network topologies: star topology, peer-to-peer topology, and cluster tree topology (see Figures 9.15 and 9.16). FFDs can be part of any topology; however RFDs are limited to a star topology.

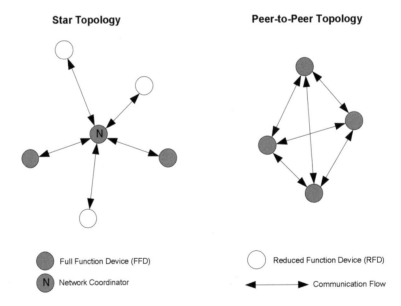

Figure 9.15 Star and Peer-to-Peer Topologies

In a star topology data may be exchanged only between the network coordinator and the DEVs or network nodes (no direct communication between DEVs). There are two types of communication: uplink, from the DEVs to the network coordinator; and downlink, from the network coordinator to the DEVs.

In a peer-to-peer topology, each DEV can communicate with others within its radio range. There is no network coordinator. However, a DEV can become such a coordinator if it is the first device to communicate on the channel. A basic peer-to-peer network can be formed when a new DEV searches for another DEV with which it can communicate. If no such DEVs are found, the new DEV can become the network coordinator and wait for other DEVs to join.

A more complex topology that can be constructed out of the peer-to-peer topology is a cluster tree topology (Figure 9.16). In the figure, the circle area represents a cluster. Multiple clusters interconnect with each other, forming a multi-cluster network. Each cluster contains network nodes and a network node designated to be a cluster head (CH). A cluster head controls its cluster. All network nodes in a cluster are only FFDs (there are no RFDs in a cluster tree network). The cluster head of the first established cluster is called a designated device (DD). Therefore, the entire multi-cluster network will have only one DD. At the cluster formation phase, a DD forms the first cluster by assigning the cluster identifier zero and transmitting the beacon. A new network node that is not associated with any cluster receives the beacon and starts making the connection request to the cluster head (or DD in the case of the first cluster). The cluster head or DD will accept the new network node to become a member of the cluster. Several network nodes join the cluster and the cluster becomes bigger and bigger. The DD may promote a network node to be the cluster head and partition a big cluster into small clusters resulting in a multi-cluster network. A unique cluster ID (CID) is assigned to each cluster.

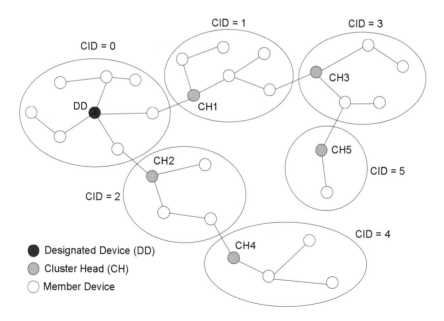

Figure 9.16 Cluster Tree Topology

9.6.1.2   Protocol Stack

The proposed standard is divided into the Link Layer Control (LLC), MAC, and its physical layer (PHY) (see Figure 9.17).

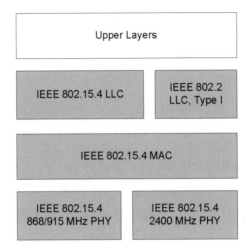

Figure 9.17 IEEE 802.15.4 Protocol Stack

9.7 IEEE 802.15.4 Physical Layer

The radio can operate at the following frequencies: the 868 MHz band (e.g., in Europe), the 915 MHz band (e.g., in the U.S.), and the 2.4 GHz band (worldwide) (Figure 9.18). At 868 MHz and 915 MHz, the transmission speed is 20 kbps using DSSS. 868 MHz band allows one channel while in the 915 MHz band there are 10 non-overlapping channels—that is, up to 10 networks can coexist in the same area. The 2.4 GHz band supports 250 kbps using DSSS, allowing 16 non-overlapping channels—that is, up to 16 networks can coexist in the same area.

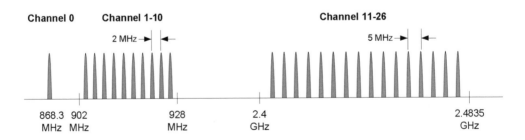

Figure 9.18 IEEE 802.15.4 Frequency Band

9.8 IEEE 802.15.4 Media Access Control

IEEE 802.15.4 focuses on low-cost, low power consumption devices. This focus constrains the MAC protocol design to be rather simple with adequate functionalities to support the low rate applications mentioned earlier. Sophisticated functions (i.e., QoS mechanisms) reside in the upper layers (above IEEE 802.15.4 MAC) which are not included in the draft standard. Several techniques are proposed to minimize the amount of overhead in maintaining the communication link. The packet structure is designed to be simple. IEEE 802.15.4 has three packet structures: beacon packet, data packet, and handshake packet (acknowledgment packet). Besides the beacon, there is no explicit message or command originated from the MAC layer that provides peer communication between the MAC layers of the two devices. The draft standard also defines several service primitives (i.e., primitives between the LLC and MAC sublayers and primitives between the MAC sublayer and the PHY) via the service access point (SAP) as shown in Figure 9.19. For more details on the concept of service primitives, please refer to IEEE 802.2. The MAC sublayer PAN information base (MAC PIB) is the MAC's database which maintains the network configuration. The peer communication between the MAC sublayers of the two devices takes place indirectly through the upper layers and the service primitives as shown in Figure 9.20.

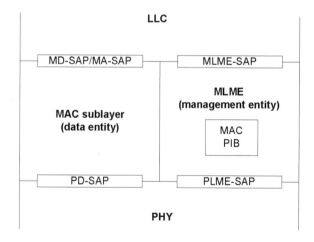

Figure 9.19 MAC Sublayer Reference Model

The proposed MAC utilizes either CSMA/CA or TDMA protocols, or a combination of both. As shown in Figure 9.21, the channel is organized into superframes. Each superframe starts with a network beacon, sent by the network coordinator. The superframe includes a beacon, a contention period, and guaranteed time slots (GTSs). The superframe duration varies depending on the active applications. If there are no low-latency applications, the superframe consists of only the beacon and the contention period. On the other hand, if there are low-latency applications, GTSs are present.

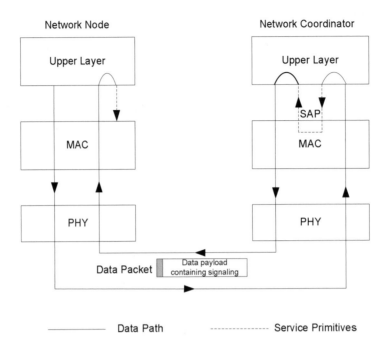

Figure 9.20 Example of Signaling Flow

In the beacon, the network coordinator 1) synchronizes the DEVs, 2) describes the structure of the superframe, and 3) notifies the pending node messages.

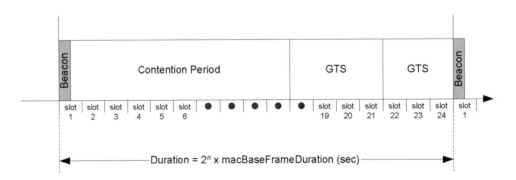

Figure 9.21 IEEE 802.15.4 Frame Structure

In the contention period, the DEVs can access the wireless media using the CSMA/CA mechanism. The CSMA/CA mechanism is similar to the one presented for the IEEE 802.11 standard (see Chapter 4). Each time a DEV wishes to transmit packets, it needs to determine if the wireless medium is free. For this process, the MAC sends a clear channel assessment (CCA) request to the physical layer. If the wireless medium is free, the packet is transmitted. If the wireless medium is busy, the DEV will back off for a random period before trying again. All data transmissions require a handshake or acknowledgment by the receiving DEV. If an acknowledgment is not received at the sender within a pre-determined time, the sender will retransmit the packet. The DEVs must stop competing for channel access at the end of the contention period.

The GTS is reserved for specific DEVs that need guaranteed bandwidth. The network coordinator may decide not to allocate a GTS to a requesting DEV or to de-allocate an existing GTS at any time.

The proposal also defines the following services required for LR-WPAN operation: network discovery by the network coordinator and DEV, network initiation by the network coordinator, network synchronization by the DEV, and network searching by the DEV. Network discovery is the process of finding out which networks exist close to the network administrator in order to be able to choose a unique network identifier and channel. Network initiation is the process of establishing the network and its operation. Network synchronization is the process of listening to the network beacon to find out the availability of communication opportunities. Network searching is the process of finding the network beacon when network synchronization has been lost.

9.9 IEEE 802.15.4 QoS Support

IEEE 802.15.4 envisions three traffic types in the LR-WPAN:

- *Periodic data:* The traffic is generated in a regular fashion. The amount of data is defined by the application itself and mostly is low data rate. An example is sensor traffic.
- *Intermittent data:* The traffic is generated once in a while, not continuously. The data generation is activated by external stimulus. An example is the light switch traffic.
- *Repetitive low-latency data:* The traffic is generated continuously and requires low-latency data transfer. An example is the mouse device traffic.

As described in the previous section, IEEE 802.15.4 includes two channel access schemes: CSMA/CA and TDMA (via GTS). CSMA/CA provides best effort service while TDMA provides quantitative service in terms of bandwidth and delay assurance. Therefore, CSMA/CA adequately supports periodic and intermittent data while TDMA is applied to repetitive low-latency data.

Repetitive low-latency data traffic requires GTSs from the network coordinator. The network coordinator checks the available network resources. If resources are available, the network coordinator reserves GTSs as well as assigns unique IDs to each GTS. The negotiation process is accomplished through the upper layers and GTS primitives (i.e., MLME-GTS.request, MLME-GTS.confirm). Figure 9.22 shows an example of GTS primitives.

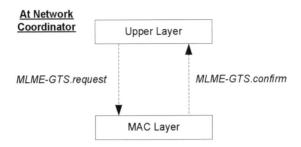

Figure 9.22 Examples of GTS Primitives

Figure 9.23 shows the GTS establishment procedure between a network node and the network coordinator. The network node's upper layer initiates the GTS request to the network coordinator. The upper layer of the network coordinator retrieves the GTS request and signals the MAC layer through the MLME-GTS.request. The MAC layer checks the available network resources. If resources are available, the request is granted, and the MAC layer replies with MLME-GTS.confirm which contains the granted GTS information (i.e., GTS ID, GTS starting time,

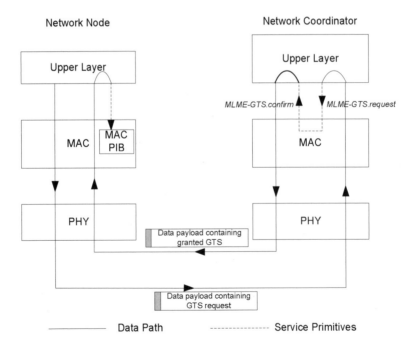

Figure 9.23 GTS Establishment Procedure

GTS length). The network coordinator's upper layer includes granted GTS information in the data payload and sends it to the network node. MLME-GTS.confirm contains GTSId, GTSStartSlot, and GTSLength. GTSStartSlot indicates the starting slot of the allocated GTS. After receiving the granted GTS information, the network node's upper layer updates its MAC PIB with GTS information. The network node accesses the channel during a GTS based on the information provided in the MAC PIB. The upper layers' mechanisms are not specified by IEEE 802.15.4.

In summary, IEEE 802.15.4 provides basic QoS mechanisms to support low-rate applications. However, the sophisticated QoS mechanisms and QoS algorithms reside in the upper layer and are not within the scope of the standard.

2.5G and 3G Networks

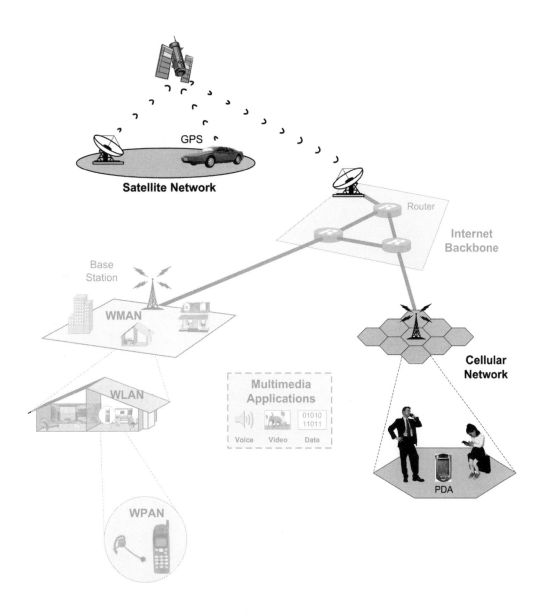

GPRS

10.1 Introduction

General Packet Radio Service (GPRS) is referred to by many as a 2.5G technology—an evolution from 2G Global System for Mobile Communications (GSM) technology and an interim phase toward 3G high-speed services including multimedia traffic with different QoS requirements.

GPRS standardization efforts were conducted by the European Telecommunications Standards Institute (ETSI) during the mid and late 1990s and then transferred to the 3G Partnership Project (3GPP) organization. 3GPP includes many organizations such as the Association of Radio Industries and Businesses (ARIB) in Japan, the China Wireless Telecommunication Standard Group (CWTS), the European Telecommunication Standard Group (ETSI), Standard Committee T1 Telecommunications (T1) in the U.S., and Telecommunication Technology Association (TTA) in Korea.

ETSI, and later 3GPP, introduced several releases of GPRS. In each release new features as well as increased speeds have been specified. Each release also has several versions in which problems are clarified. Each release and version is detailed in several documents that provide the release overview as well as detailed specifications of its various modules and parts. 3GPP started to harmonize later releases of GPRS to enable easy migration to the anticipated 3G Universal Mobile Telecommunications System (UMTS) technology (described in Chapter 11), which is provisioned by Wide Code Division Multiple Access (WCDMA) technology. The evolution of these standards is presented in Figure 10.1. It is expected that most of the telephone carriers will take this migration path, or part of it, toward 3G. This includes both carriers that currently support GSM and carriers that support alternative technologies such as CDMA. Such CDMA carriers are located mainly in the U.S. and Korea.

GSM, which was introduced in the early 1990s, is available in more than 100 countries and is the de facto standard in Asia and Europe. GSM is based on narrowband Time Division Multiple Access (TDMA), which can accommodate eight simultaneous calls on the same frequency for circuit switching services. Its standards are specified by ETSI and the TIA/EIA (Telecommunications Industry Alliance/Electronics Industries Alliance) IS-136 standards. GSM charges are based on the connection time and not on the volume of traffic delivered. This cost model is not suitable for certain data applications that are characterized by bursty transmissions.

GSM Phase 1 included common services such as basic telephony, emergency calls, up to 9.6 kbps data rate services, ciphering, authentication, and features such as call forwarding and SMS (Short Message Service). GSM Phase 2 included additional features such as identification, call waiting, call hold, advice of charge, and multi-party call. A half-rate speech coding was introduced in this phase in addition to full-rate speech coding. These two phases provided the basis for work toward 3G, referred to also as Phase 2+, which includes GPRS specifications.

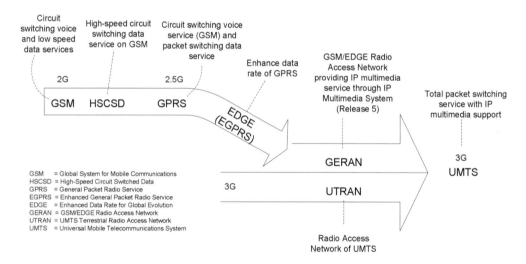

Figure 10.1 Evolution toward 3G/UMTS

The first work under Phase 2+, referred to as Release 96 (Rel-96), specified HSCSD (High-Speed Circuit Switched Data). HSCSD increased the 9.6 kbps data rate services provided in GSM Phases 1 and 2 to a maximum of 115 kbps. This data rate is achieved through a relatively simple combination of multiple GSM time slot allocations. However, due to various implementations and interfaces, the practical data rate is up to 64 kbps. In HSCSD, the dedicated time slots that are reserved for data connections cause inefficient bandwidth utilization in cases where the data traffic is bursty. GPRS releases continued to support HSCSD, which is currently deployed only in a few locations.

GPRS evolves on top of GSM's TDMA approach. The first release of GPRS was detailed in Release 97 (R97). R97 introduced the concept of packet switching in addition to GSM's circuit switching. The packet switching allows carriers to charge users based on actual traffic transmitted rather than on the duration of the circuit. GPRS also allows users to access other public networks via protocols such as IP. R97 provides speeds of up to 171 kbps with average speeds of 28.8 kbps. R97 allows support for QoS considering service priority, reliability, time delay, throughput, QoS profile, and other parameters. Release 98 (R98) included Adaptive Multi Rate (AMR) technology, which provides the ability to select either full-rate (FR) or half-rate (HR) speech coding. These releases are being deployed globally despite their relatively low data rates.

Release 99 (R99) includes Enhanced Data rate for Global Evolution (EDGE), sometimes referred to as Enhanced Data GSM Environment and as Enhanced GPRS (EGPRS). EDGE provided a number of additional radio modulation and coding techniques that enhanced speeds of up to 384 kbps with average speeds of 64 kbps. This release includes EDGE Compact, which is a variant of EGPRS that was designed to be deployed in narrow frequency band allocations. EDGE Compact requires less than 1MHz of spectrum, compared to more than 2.4 MHz for EDGE, reducing EDGE bandwidth to up to 250 kbps with averages of up to 56 kbps.

R99 also included support for UMTS traffic. This inclusion of UMTS traffic has been continuously improved and enhanced through the Release 4 (Rel-4), temporarily referred to as Release 2000, and Release 5 (Rel-5), known also as GSM/EDGE Radio Access Network (GERAN). In GERAN a new radio access technology was introduced that is fully harmonized with UMTS Terrestrial Radio Access Network (UTRAN), which details speeds of up to 2 Mbps. Rel-5 includes an optional Enhanced GPRS (EGPRS) radio that allows enhanced speeds compared to GPRS. Rel-5 also includes a new set of multimedia services, referred to as Internet Multimedia Subsystem (IMS), including multimedia interactive communications.

3GPP continues to work on a newer release, referred to as Release 6 (Rel-6). Rel-6 may include improved radio technology for increased data rates and other improvements. In this book we focus on Rel-5 while, when possible, we provide some pointers to previous releases. These intermediate releases receive interest from various carriers that may deploy them as an intermediate step toward 3G technology.

GPRS Release 5 is specified in several 3GPP documents including:

- TS 22.060 3rd Generation Partnership Project; Technical Specification Group Services and System Aspects; General Packet Radio Service (GPRS); Service description, Stage 1 (Release 5)
- TR 22.941 3rd Generation Partnership Project; Technical Specification Group Services and System Aspects; IP Based Multimedia Services Framework; Stage 0 (Release 5)
- TS 22.228 3rd Generation Partnership Project; Technical Specification Group Services and System Aspects; Service requirements for the IP Multimedia Core Network Subsystem (Stage 1) (Release 5)

- TS 43.051 3rd Generation Partnership Project; Technical Specification Group GSM/EDGE Radio Access Network; Overall description (Stage 2) (Release 5)
- TS 44.118 3rd Generation Partnership Project; Technical Specification Group GSM EDGE Radio Access Network; Mobile radio interface layer 3 specification, Radio Resource Control (RRC) Protocol, Iu Mode (Release 5)
- TS 44.018 3rd Generation Partnership Project; Technical Specification Group GSM/EDGE Radio Access Network; Mobile radio interface layer 3 specification; Radio Resource Control Protocol (Release 5)
- TS 44.060 3rd Generation Partnership Project; Technical Specification Group GSM/EDGE Radio Access Network; General Packet Radio Service (GPRS); Mobile Station (MS)—Base Station System (BSS) interface; Radio Link Control/Medium Access Control (RLC/MAC) protocol (Release 5)
- TS 44.160 3rd Generation Partnership Project; Technical Specification Group GSM/EDGE Radio Access Network; Mobile Station (MS)—Base Station System (BSS) interface; Radio Link Control/Medium Access Control (RLC/MAC) protocol; Iu mode (Release 5)
- TS 43.064 3rd Generation Partnership Project; Technical Specification Group GERAN; Digital cellular telecommunications system (Phase 2+); General Packet Radio Service (GPRS); Overall description of the GPRS radio interface (Stage 2) (Release 5)
- TS 45.005 3rd Generation Partnership Project; Technical Specification Group GSM/EDGE Radio Access Network; Radio transmission and reception (Release 5)

Several other documents detail various aspects of the standards. They can be found at www.3gpp.org and www.etsi.org.

10.2 GPRS (Rel-5) Architecture

As shown in Figure 10.2, the GPRS mobile system consists of three main entities: Mobile Station (MS), Radio Access Network (RAN), and Core Network (CN). A mobile station refers to the mobile equipment on which a user executes mobile services and applications (i.e., voice services, SMS). A mobile station communicates with the base station that controls the radio channel within its coverage area (or a cell, so to speak). Multiple base station systems interconnect and form the network called RAN. RAN is responsible for providing the communication path between the Mobile Station and the Core Network. Furthermore, RAN also provides several key functions to support communication such as channel access, mobility management, and radio resource management. GERAN (GSM/EDGE Radio Access Network) is the RAN entity of GPRS (Rel-5).

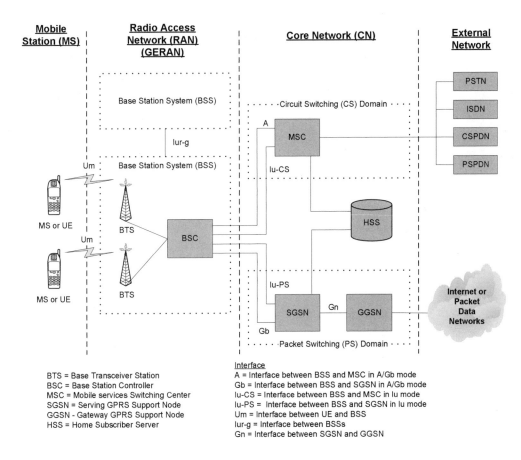

Figure 10.2 GPRS Architecture

The Core Network connects the mobile network to the external network (i.e., [PSTN], Integrated Services Digital Network [ISDN], Internet). The core network comprises two domains: Circuit Switching (CS) domain and Packet Switching (PS) domain. The CS domain contains modules that support circuit switching services. As shown in Figure 10.2, the Mobile Switching Center (MSC) resides in the CS domain. The PS domain contains the following main modules that support packet switching services: Serving GPRS Support Node (SGSN) and Gateway GPRS Support Node (GGSN). A voice signal (circuit switching traffic) travels between a mobile station and an external voice network (i.e., PSTN) through the Base Station System (BSS) and MSC. On the other hand, a data signal (packet switching traffic) travels between a mobile station and an external data network (i.e., Internet) through the BSS, SGSN, and GGSN. Home Subscriber Server (HSS) is a supporting module providing subscription-related information for both the CS and PS domains.

GERAN consists of multiple interconnected BSSs. Each BSS serves the mobile stations within its coverage area. The BSS connects to the CS domain of the core network through A and Iu-CS interfaces while the BSS connects to the PS domain of the core network through Gb and Iu-PS interfaces. Iu-CS and Iu-PS can be simply called Iu interface. The mobile station connects to the BSS through the Um interface. Each interface has its own protocol architecture that handles the communication between the pair systems.

Prior to Release 5, the BSS connects to the core network only via the A and Gb interfaces (no Iu interface). This core network is considered to be the second generation core network. 3GPP realizes a smooth migration from the second generation mobile system toward 3G's UMTS (Universal Mobile Telecommunications System). In UMTS, UTRAN (UMTS Terrestrial Radio Access Network) connects to the third generation core network through the Iu interface. Therefore, in order to align with UMTS, GPRS Release 5 includes Iu interface in addition to the existing A/Gb interfaces. Figure 10.3 illustrates the GPRS architecture for 2G and 3G core networks.

The mobile station can be in one of two modes: A/Gb mode and Iu mode. The mode is determined by the generation of the core network (i.e., 2G or 3G) with which the mobile station associates (see Figure 10.3).

The standard allows three domains of operation for the mobile station in A/Gb mode:

- *Class A domain of operation:* The MS can operate both packet switching services and circuit switching services simultaneously.
- *Class B domain of operation:* The MS can operate either packet switching services or circuit switching service at one time.
- *Class C domain of operation:* The MS can only operate packet switching services.

Furthermore, the standard also defines three domains of operation for the mobile station in Iu mode:

- *CS/PS domain of operation:* The MS can operate both packet switching services and circuit switching services simultaneously. This domain is analogous to Class A of the A/Gb mode.
- *PS domain of operation:* The MS can only operate packet switching services. This domain is analogous to Class C of the A/Gb mode
- *CS domain of operation:* The MS can only operate circuit switching services.

GPRS supports both Point-to-Point (PTP) and Point-to-Multipoint (PTM) services while supporting roaming of GPRS users between different carriers. The PTP can be used for accessing information on the Internet, messaging services, and conferencing applications. PTM can be used for delivering information, such as news and weather, to multiple locations or interactive conferencing applications.

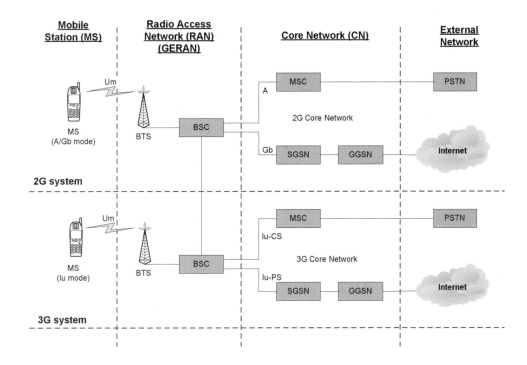

Figure 10.3 GPRS Architecture for 2G and 3G Systems

GPRS Rel-5 added a new component as compared to previous GPRS releases, the Internet Multimedia Subsystem (IMS). IMS allows simultaneous access to multiple different types of real-time and non-real-time traffic. Real-time traffic includes voice, text, and video, and non-real-time traffic includes data, audio and video files, audio and voice streaming, text and multimedia messaging, emails, and Internet browsing.

IMS provides synchronization between the multiple components of a multimedia communication session. For example, a user can receive calls while continuing with web browsing. IMS lists key requirements for multimedia services. These requirements pertain to the user perspective. In addition, IMS lists requirements for security, addressing, support for roaming, emergency calls, mobile number portability, messaging, and Virtual Private Network (VPN) support.

10.2.1 Radio Interface Protocol Architecture

Figure 10.4 illustrates the BSS's radio interface protocol stack, which includes the protocol modules for Iu and A/Gb modes. The protocol architecture is divided into 1) user plane for user data transfer and the associated user data transfer control and 2) control plane for the transfer of control information that supports the user plane.

A logical channel that represents a traffic stream of a specific type of information (i.e., data message, control message) contains a specific structure of the information block (or packet) which is mapped to the physical layer. The logical channel will be described in a later section. The radio interface protocol is divided into three layers: Layer 1 or physical layer (PHY); Layer 2, which includes the Data Link, Media Access Control (MAC), Radio Link Control (RLC) and Packet Data Convergence Protocol (PDCP); and Layer 3, which contains the Radio Resource Control (RRC) for Iu mode and Radio Resource (RR) for A/Gb mode.

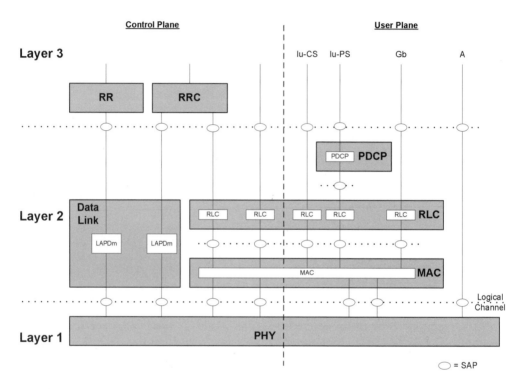

Figure 10.4 BSS's Radio Interface Protocol Stack

10.2.1.1   RRC

The Radio Resource Control (RRC), introduced in Rel-5, is responsible for radio resource management in the Iu mode. In these matters it shares similar responsibilities with the MAC. The RRC will carry the resource management responsibilities when dedicated resources are provided to the mobile station, where as the MAC will be responsible when shared resources are provided to the mobile station. The RRC broadcasts system information to the mobile stations. The system information may be cell specific (such as location services in cells) or not cell specific. The RRC is also responsible to establish, maintain, reconfigure, and release connections between the mobile stations and the GERAN. This includes cell reselection, admission control, and link establishment. The RRC selects the connection parameters considering both control and applications needs. This allocation is then communicated to the mobile station. The RRC also handles mobility related activities such as evaluation, decision, and execution handovers. The RRC handover procedures follow the mobility of the mobile station and include procedures to modify the channels allocated to the mobile station. While executing these functions, the RRC considers the requested QoS. In addition, the RRC executes a variety of monitoring, reporting, ciphering, and mediating between GERAN and the mobile stations.

10.2.1.2   RR

The Radio Resource (RR) is responsible for the resource management in A/Gb mode. The RR is located in a plane parallel to the RRC.

10.2.1.3   PDCP

The Packet Data Convergence Protocol (PDCP) is located above the RLC layer in the user plane of the Iu mode. PDCP includes the following functions:

- Header compression and decompression of the IP data stream: PDCP receives the packet from the upper layer, performs header compression, and forwards the packet to the RLC layer. In the opposite direction, PDCP receives the packet from the RLC layer, decompresses the header, and forwards the packet to the upper layer.
- Passage of the data between the upper and lower layers.
- Maintenance of the PDCP sequence numbers: PDCP assigns a sequence number to the packet that arrives from the upper layer. At the receiving end, PDCP receives the packet with the sequence number from the RLC layer. Based on the sequence number it decides its action.

10.2.1.4   RLC

The RLC transmits data from the higher layers to the MAC. It can work in either transparent or non-transparent mode. In transparent mode the RLC has no functionality. Packets are transferred to the MAC without being altered and without adding any RLC protocol information. In non-transparent mode, the RLC provides reliability to the MAC data transmission. To achieve this, the RLC appends protocol information (RLC packet header) to the data packet and performs functions such as 1) segmentation of upper layer packets into RLC data blocks, 2) reassembly of RLC data blocks and transfer of the blocks back to the upper layer, 3) concatenation of upper layer packets into RLC data blocks, 4) padding to fill RLC data blocks, 5) backward Error Correction (BEC) procedures enabling selective retransmission of RLC data blocks, 6) discarding of RLC packets according to delay requirements, 7) in-sequence delivery of upper layer packets, 8) Link Adaptation, 9) Ciphering, and 10) validating Acknowledgments (if desired by upper layers). The non-transparent mode can be divided into Acknowledged mode and Unacknowledged mode. Acknowledgment mode provides a mechanism for guaranteed delivery since the addressee needs to acknowledge the receipt of the packet. The RLC can notify the upper layers when errors cannot be resolved as well as when packets are discarded. For unacknowledged mode, RLC does not provide guaranteed delivery of packets.

10.2.1.5   MAC

The Medium Access Control (MAC) executes the functions of managing the shared transmission resources. In these matters it shares similar responsibilities with the RLC. The MAC responsibilities include 1) configuring and mapping logical channels, defined by the type of data which is transferred, to the appropriate physical subchannels; 2) assigning and configuring the shared radio resources; and 3) scheduling, multiplexing data, and controlling packets to be delivered on the shared physical channels. The MAC supports various logical Channels: traffic channels (TCH), control channels, and broadcast channels. The MAC supports both Iu and Gb traffic.

10.2.1.6   Data Link

The LAPDm delivers information between Layer 3 and Layer 1 across the GSM network. The term Dm channel is used to represent the collection of all the various signaling channels required in the GSM system.

10.2.1.7   Physical Layer

The physical channel interfaces with the Medium Access Control (MAC) and the Radio Resource Control (RRC). It uses a combination of frequency and Time Division Multiple Access (TDMA) methods. Each frequency channel is separated by 200 kHz. In each frequency channel, the channel is structured into time slots and defines eight basic physical channels per carrier. The physical channel is defined as a sequence of time slots and a frequency hopping sequence. Several frequencies are defined for usage in different countries. According to 3GPP TS 45.001 document, the frequencies and channel arrangements are as follows:

1. GSM 450 Band: the system is required to operate in the following band:

    a. 450.4 MHz to 457.6 MHz: mobile transmit, base receive
    b. 460.4 MHz to 467.6 MHz: base transmit, mobile receive

2. GSM 480 Band: the system is required to operate in the following band:

    a. 478.8 MHz to 486 MHz: mobile transmit, base receive
    b. 488.8 MHz to 496 MHz: base transmit, mobile receive

3. GSM 750 Band: the system is required to operate in the following band:

    a. 777 MHz to 792 MHz: mobile transmit, base receive
    b. 747 MHz to 762 MHz: base transmit, mobile receive

4. GSM 850 Band: the system is required to operate in the following band:

    a. 824 MHz to 849 MHz: mobile transmit, base receive
    b. 869 MHz to 894 MHz: base transmit, mobile receive

5. Standard or primary GSM 900 Band, P-GSM: the system is required to operate in the following frequency band:

a. 890 MHz to 915 MHz: mobile transmit, base receive
b. 935 MHz to 960 MHz: base transmit, mobile receive

6. Extended GSM 900 Band, E-GSM (includes Standard GSM 900 band): the system is required to operate in the following frequency band:

a. 880 MHz to 915 MHz: mobile transmit, base receive
b. 925 MHz to 960 MHz: base transmit, mobile receive

7. Railways GSM 900 Band, R-GSM (includes Standard and Extended GSM 900 Band): the system is required to operate in the following frequency band:

a. 876 MHz to 915 MHz: mobile transmit, base receive
b. 921 MHz to 960 MHz: base transmit, mobile receive

8. DCS 1800 Band: the system is required to operate in the following band:

a. 1710 MHz to 1785 MHz: mobile transmit, base receive
b. 1805 MHz to 1880 MHz: base transmit, mobile receive

9. PCS 1900 Band: the system is required to operate in the following band:

a. 1850 MHz to 1910 MHz: mobile transmit, base receive
b. 1930 MHz to 1990 MHz base transmit, mobile receive

To carry its transmission responsibilities successfully, the physical layer is also responsible for power control. The physical layer adjusts the mobile station power output level such that the desired quality is achieved with the minimum possible power. It synchronizes the receiving station with regard to frequency and time. It also carries handover functionality when a mobile station moves from one cell to another as well as quality monitoring that may result in a station changing its assigned physical channel.

The channel modulation can be either GMSK or 8-PSK. Each modulation can achieve a different data rate. The four channel coding schemes, CS-1 to CS-4, are defined for GPRS, while nine channel coding schemes, MCS-1 to MCS-9, are defined for EGPRS (EDGE). The bandwidth allocated for speech or data transmission is provided by a combination of modulation, block size, half or full rate, code rate, and other radio parameters. The potential speed of speech, control, and data varies significantly. Speech varies up to around 24 kbps, data varies up to around 480 kbps.

10.2.2 GPRS (Rel-5) Protocol Architecture

Figures 10.5 and 10.6 show the user and control plane protocols, respectively, in the PS domain (both A/Gb mode and Iu mode). Figures 10.7 and 10.8 show the user and control plane protocols, respectively, in the CS domain (both A/Gb mode and Iu mode). In these figures we do not intend to include all the protocols' components and connections, but rather to provide the big picture of the GPRS (Rel-5) protocol architecture.

As shown in Figure 10.5, the functions between A/Gb mode and Iu mode are slightly different. Iu RLC includes the transparent mode in addition to the acknowledged mode and un-acknowledged mode in A/Gb RLC. The PDCP used in Iu mode and the Sub-Network Dependent Convergence Protocol (SNDCP) used in A/Gb mode provide the logical link control between the MS and the BSS. In Figure 10.5, the RR operates the radio resource management in A/Gb mode while the RRC operates the radio resource management in Iu mode. Other modules include the Radio Access Network Application Part (RANAP), the Signaling Connection Control Part (SCCP), and the Base Station System GPRS Protocol (BSSGP). These modules are not discussed in this book.

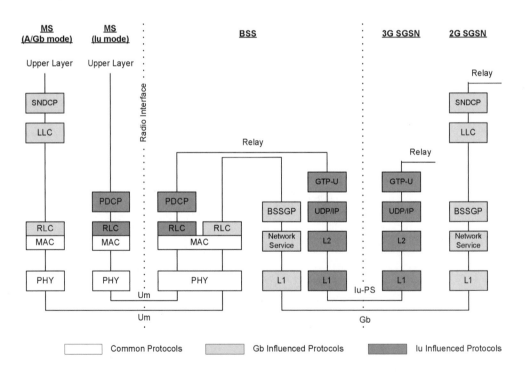

Figure 10.5 User Plane Protocol Architecture in PS Domain (A/Gb mode and Iu mode)

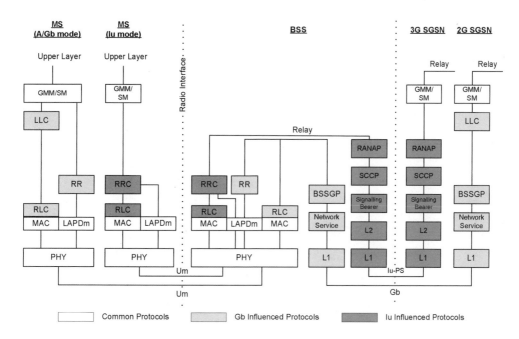

Figure 10.6 Control Plane Protocol Architecture in PS Domain (A/Gb mode and Iu mode)

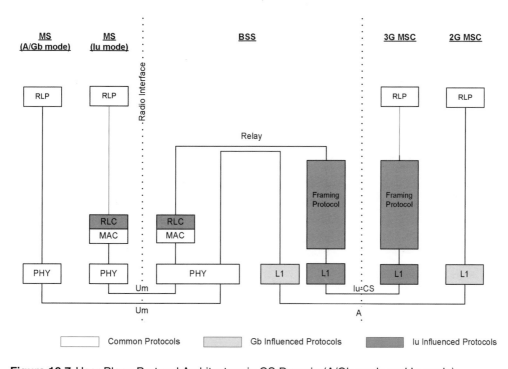

Figure 10.7 User Plane Protocol Architecture in CS Domain (A/Gb mode and Iu mode)

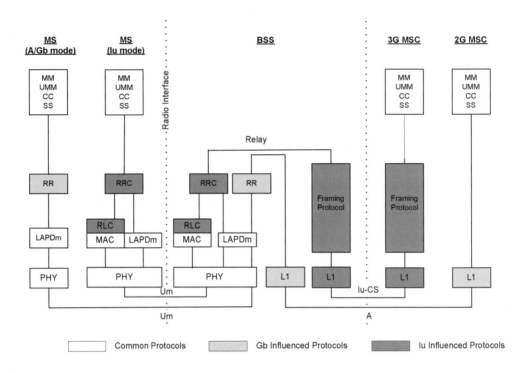

Figure 10.8 Control Plane Protocol Architecture in CS Domain (A/Gb mode and Iu mode)

10.3 Physical Channel

The physical channel is defined by a combination of a radio frequency channel and a time slot number. Within the frequency band, a radio frequency channel (seperated by 200 kHz) is structured into time slots (TSs) as shown in Figure 10.9. Eight consecutive time slots form a TDMA frame. A time slot in a TDMA frame is numbered from 0 to 7. A physical channel is defined by a radio frequency channel and a time slot number (TN) in every TDMA frame. All downlink and uplink TDMA frames are aligned. At the BSS, an uplink TDMA frame is delayed by three time slots from a downlink TDMA frame. At the mobile station, this delay is a function of the distance (the propagation delay) between the mobile station and the base station.

A physical channel can be categorized as a Shared Basic Physical Sub Channel (SBPSCH) or a Dedicated Basic Physical Sub Channel (DBPSCH). A DBPSCH is dedicated to only one user. In contrast, SBPSCH can be shared by up to eight users. Uplink Stage Flag (USF) is deployed to control the multiple access of users on SBPSCH. We will describe the USF mechanism in Section 10.5.2. Figure 10.10 shows an example of the physical channel assignment.

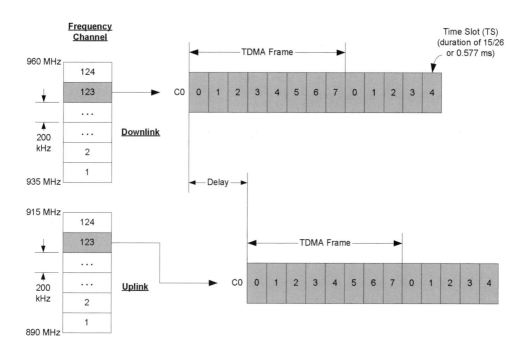

Figure 10.9 Physical Channel Structure of Standard GSM 900 Band

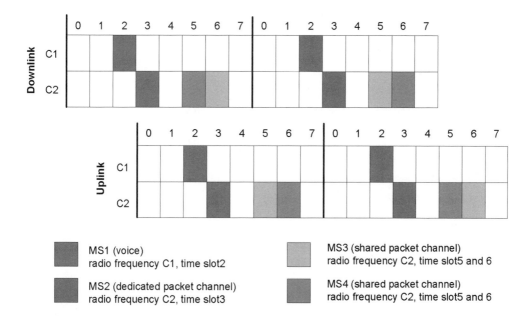

MS1 (voice)
radio frequency C1, time slot2

MS3 (shared packet channel)
radio frequency C2, time slot5 and 6

MS2 (dedicated packet channel)
radio frequency C2, time slot3

MS4 (shared packet channel)
radio frequency C2, time slot5 and 6

Figure 10.10 Example of Physical Channel Assignment

10.3.1 Hyperframes, Superframes, Multiframes, TDMA Frames and Time Slots

As shown in Figure 10.11, the standard defines different duration frames. The longest frame, called *hyperframe*, with a duration of 12533.76 seconds, is defined to support cryptographic mechanisms. A hyperframe is divided into 2048 *superframes*, each with duration of 6.12 seconds. A superframe is divided into *multiframes*. The number of multiframes in a superframe is based on the type of multiframes. There are four types of multiframes:

- 26-Multiframe: This is a 26-multiframe that contains 26 TDMA frames. Therefore, a superframe can accommodate 51 26-multiframes.
- 51-Multiframe: This is a 51-multiframe that contains 51 TDMA frames. Therefore, a superframe can accommodate 26 51-multiframes.
- 52-Multiframe: This 52-multiframe, which is constructed from two 26-multiframes, contains 52 TDMA frames. Therefore, a superframe can accommodate 26 52-multiframes.
- 52-Multiframe for CTS (Cordless Telephone System): This multiframe contains 52 TDMA frames.

Each type of multiframe supports a different type of logical channel. Figure 10.11 shows only the 26-multiframe type. However, we will focus more on the 52-multiframe type, which is used to accommodate data and control packets.

A TDMA frame contains eight time slots, each with a duration of 0.577 ms. A time slot can accommodate a burst size of 156.25 symbols. If GMSK modulation scheme is used, one symbol reflects one bit (i.e., a time slot contains 156.25 bits). If an 8-PSK modulation scheme is used, one symbol reflects three bits (i.e., a time slot contains 469.75 bits).

10.3.2 Packet Data Channel (PDCH)

The Packet Data Channel (PDCH) refers to the physical channel dedicated to packet data traffic—that is, control or user data traffic from the logical channel. PDCH uses 52-multiframe structure. As shown in Figure 10.12, a PDCH multiframe comprises 52 TDMA frames which are divided into: 12 blocks (four TDMA frames per block), two idle frames, and two frames used for PTCCH (Packet Timing Advance Control Channel).

A PDCH is defined by a combination of a radio frequency channel and a time slot. On a radio frequency channel, PDCHs can be represented logically as eight physical channel instances (see Figure 10.13). The PDCH number presented here corresponds to the time slot number of the TDMA frame (i.e., PDCH1 and time slot # 1)

10.3.3 Radio Block

The radio block is defined as a sequence of data bursts that contains one RLC/MAC protocol data unit (PDU). A packet is generated at the application layer and is sent to the lower layer. Each layer appends to the packet its own protocol header and trailer. Figure 10.14 illustrates the

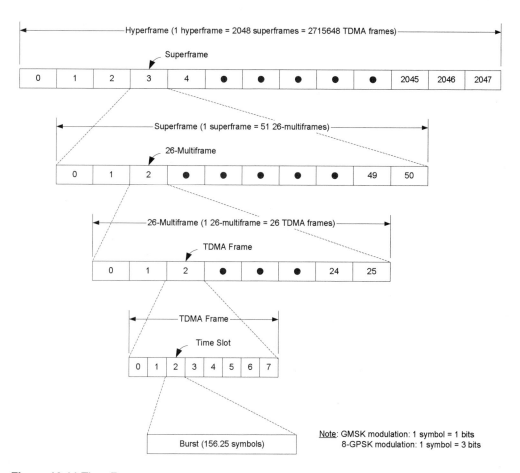

Figure 10.11 Time Frame

packet flow and its structure in each protocol layer. It is worth mentioning that the RLC/MAC PDU header contains the Uplink State Flag (USF) defined above, which plays an important role in the dynamic channel assignment. At the physical layer, the RLC/MAC PDU will be coded (using a coding scheme such as CS-1 to CS-4, or MCS-1 to MCS-9) and reformatted to fit into a PDCH block.

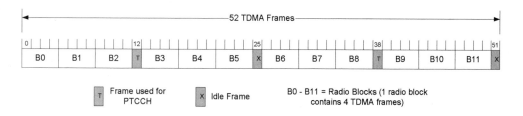

Figure 10.12 52-Multiframe for PDCHs

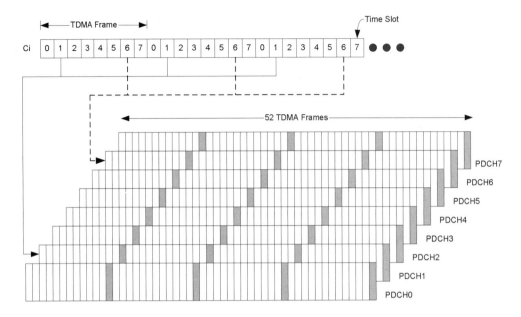

Figure 10.13 PDCHs' Logical View

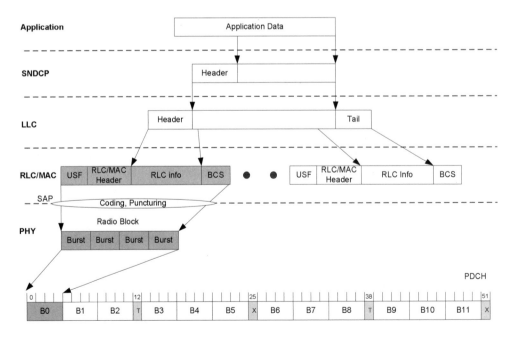

Figure 10.14 Packet Flow and Structure

10.4 Logical, Control, and Traffic Channels

10.4.1 Logical Channels

The standard defines logical channels that are mapped by the MAC to physical channels. These logical channels include both traffic channels for data and speech and control channels for control, synchronization, and signaling information. Several logical channels can be multiplexed on the downlink physical channel. Similarly, several logical channels can be multiplexed on the uplink physical channel. Since GPRS is built on the basis of GSM, some logical channels are inherited from GSM. In addition, GPRS introduced some new logical channels to support packet switching services. As shown in Figure 10.15, the naming of such logical channels begins with the word "packet," such as Packet Common Control Channel (PCCCH) and Packet Broadcast Control Channel (PBCCH).

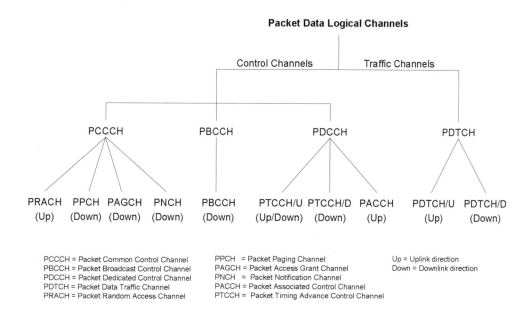

Figure 10.15 Packet Data Logical Channels

10.4.2 Control Channels

Control channels carry control, synchronization, and signaling information. There are three kinds of control channels: common, dedicated, and broadcast.

Common control channels include also Packet Common Control Channels (PCCCHs) that support packet transmission. They include packet paging channels, random access channels, packet grant channels, and packet notification channels. PCCCH is optional for the network. In case PCCCH is not allocated, the control information will be transmitted on the GSM control channels.

Dedicated control channels support operations on DBPSCH. They include Slow Associated Control Channel (SACCH) for radio measurements and data. SACCH is also used for SMS transfer during calls. They also include Fast Associated Control Channel (FACCH) for one TCH on DBPSCH and Stand-alone Dedicated Control Channel (SDCCH). Control channels supporting SBPSCH include controlling uplink and downlink activities using the Packet Associated Control Channel (PACCH), and Packet Timing Advance Control Channels (PTCCHs) for timing estimation.

Broadcast channels include 1) frequency correction channels that are used to correct the frequency of the mobile station (MS), 2) synchronization channels that synchronize the MS frequency with that of the base station, 3) broadcast control channels that broadcast general information on the base station, and 4) packet broadcast channels that broadcast parameters that the MS needs in order to access the network for packet transmission.

The control signaling messages include the following:

1. Packet Random Access Channel (PRACH) and Compact Packet Random Access Channel (CPRACH) are used by the mobile station in the uplink channel to initiate data or control information.
2. Packet Paging Channel (PPCH) and Compact Packet Paging Channel (CPPCH) are used to page a mobile station in the downlink channel before data transmission.
3. Packet Access Grant Channel (PAGCH) and Compact Packet Access Grant Channel (CPAGCH) are used in the downlink channel to transmit to the mobile station resource assignments as a preparation step toward data transmission.
4. Packet Notification Channel (PNCH) and Compact Packet Notification Channel (CPNCH) are used in the downlink to send a PTM-M (Point to Multipoint-Multicast) notification to several mobile stations about a data transfer.
5. Packet Broadcast Control Channel (PBCCH) and Compact Packet Broadcast Control Channel (CPBCCH) broadcast packet data specific system information in the downlink.
6. Packet Associated Control Channel (PACCH) is used to provide signaling information of a specified mobile station such as acknowledgments and power control information. It is also used to transmit resource assignment and reassignment messages.
7. Packet Timing Advance Control Channel/Uplink (PTCCH/U) is used to transmit random access burst for estimation of the timing advance for the mobile station in packet transfer mode.
8. Packet Timing Advance Control Channel/Downlink (PTCCH/D) is used to transmit timing advance information updates to several mobile stations. One PTCCH/D is paired with several PTCCH/Us.
9. Broadcast Control Channel (BCCH) in the downlink is used to broadcast cell specific information.

10.4.3 Packet Traffic Channels

Traffic channels (TCHs) carry either encoded speech or user data. These channels are multiplexed in either a predetermined manner or dynamically by the MAC. TCH can be either full rate (TCH/F) or half rate (TCH/H). Traffic channels are also distinguished by the modulation technique used such as GMSK and 8-PSK. Packet Data Traffic Channel (PDTCH) is for data transfer on both physical channels SBPSCH (Shared Basic Physical Sub Channel and DBPSCH (Dedicated Basic Physical Sub Channel). The PDTCH is a temporary channel provided to a single mobile station or to a group in case of multicast transmission. However, one mobile station may use several PDTCHs in parallel. PDTCH allows several MSs to be multiplexed on the same SBPSCH. PDTCH also allows several traffic classes to be multiplexed on the same shared or dedicated channel. In a multicast transmission a mobile station may use several PDTCHs for individual packet transfer. The PDTCH is either in the uplink direction (PDTCH/U) or downlink direction (PDTCH/D). A PDTCH may be either full rate (PDTCH/F) or half rate (PDTCH/H) depending on whether it is transmitted on a Packet Data Channel/Full (PDCH/F) or Packet Data Channel/Half (PDCH/H), respectively.

10.5 Media Access Control (MAC) and Radio Link Control (RLC)

The Media Access Control (MAC) and Radio Link Control (RLC) provide data traffic transfer from the upper layer to the physical channel. The MAC is connection oriented—that is, connections need to be established before any packets can be transmitted. In the standard such connections are called Temporary Block Flows (TBFs). A TBF is a logical unidirectional connection between two MAC entities, where one MAC entity belongs to the mobile station and the other to the BSS. The TBF is temporary entity since the connection ends once there are no more data to transfer and all acknowledgments have been successfully received. The TBF is allocated radio resources on one or more physical channels (i.e., PDCHs). All packets (i.e., RLC/MAC PDUs) that belong to the TBF are transmitted on the allocated PDCHs. A PDCH can accommodate multiple TBFs. To differentiate between TBFs, the standard introduces Temporary Flow Identity (TFI). The TFI value is unique among concurrent TBFs in the same direction (uplink or downlink). The TFI, together with its direction, provides a TBF unique identifier. In addition, a Global_TFI is assigned to each station for identification purposes.

In the following subsections we will focus on some key mechanisms incorporated in the GPRS MAC.

10.5.1 TBF Establishment

10.5.1.1   TBF Establishment Initiated by Mobile Station

The mobile station which is the source of the data traffic initiates TBF establishment with the BSS using one of the following access processes: One Phase Access or Two Phase Access. These processes are described below.

10.5.1.1.1 One Phase Access   In the One Phase Access (Figure 10.16), the mobile station initiates the packet access procedure by sending PACKET CHANNEL REQUEST messages on the PRACH channel. The PACKET CHANNEL REQUEST messages contain an indication of the type of access, parameters, and requirements for radio resources. This includes the requested TBFs' type, Radio Priority, and number of radio blocks (or RLC/MAC blocks). The number of requested radio blocks cannot exceed the actual number of RLC/MAC blocks waiting in the mobile station. Due to the nature of the random access scheme used in PRACH, it is possible to have multiple stations sending PACKET CHANNEL REQUEST at the same time (i.e., collisions can occur). If a collision or packet loss occurs (i.e., the mobile station does not receive a response from the base station), the PACKET CHANNEL REQUEST will be retransmitted. The mobile station is allowed up to a predetermined number of attempts to send a PACKET CHANNEL REQUEST. The maximum number of attempts is combined with a persistence level that is in direct relationship to the mobile station priority. The higher the priority, the higher the persistence level. The number of allowed attempts is broadcasted by the base station on the PBCCH and PCCCH.

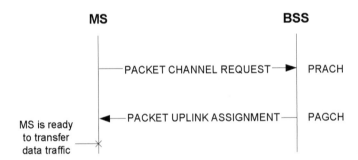

Figure 10.16 TBF Establishment Initiated by the MS: One Phase Access

After successful reception of the PACKET CHANNEL REQUEST, the BSS replies on the PAGCH channel with PACKET UPLINK ASSIGNMENT. The PACKET UPLINK ASSIGNMENT contains the following information:

- TFI: Unique value that identifies the requested TBF.
- One or more PDCHs assigned to the TBF: A radio frequency channel and one or more time slots.
- USFs: 3-Bit field associated with each assigned PDCH. For example, if a TBF connection is assigned frequency c_i and time slot numbers 2 and 5, there are two USFs assigned: one for time slot number 2 and the other for slot number 5.
- USF_GRANULARITY: 1-Bit field that indicates the number of RLC/MAC blocks transmitted at a time. Value "0" refers to one RLC/MAC block; value "1" refers to four consecutive RLC/MAC blocks.

10.5.1.1.2 Two Phase Access In the Two Phase Access (Figure 10.17), the mobile station initiates the packet access procedure by sending PACKET CHANNEL REQUEST. The BSS requests the mobile station to send a PACKET RESOURCE REQUEST message. This request is sent implicitly to the mobile station in the PACKET UPLINK ASSIGNMENT message. Upon receiving the PACKET UPLINK ASSIGNMENT, the mobile station transmits a PACKET RESOURCE REQUEST message to the BSS on PACCH. The PACKET RESOURCE REQUEST contains the following information:

- Mobile station radio access capability.
- Channel Request Description: the request uplink resource includes the radio peak throughput, priority class of the requested TBF, RLC mode of the requested TBF (i.e., RLC acknowledged mode, RLC unacknowledged mode), and the number of RLC/MAC blocks that the mobile station wishes to transmit.

After receiving the PACKET RESOURCE REQUEST, the BSS replies on the PACCH channel with PACKET UPLINK ASSIGNMENT.

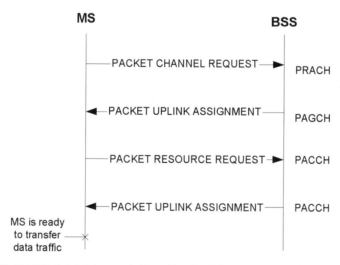

Figure 10.17 TBF Establishment Initiated by the MS: Two Phase Access

10.5.1.2 TBF Establishment Initiated by Network
The network may initiate a TBF to transfer packets from the network to the mobile station. The procedure may be entered when the mobile station is in MAC-Idle state (for Iu mode) or in Packet Idle State (for A/Gb mode) (Figure 10.18). Such TBF can also be initiated on PACCH if a TBF in this direction is already established. First the BSS performs a paging procedure to discover the location of the mobile station. After discovery, the BSS will perform resource assignment. The network may assign one or more PDCHs for the TBFs based on the network's discretion. The allocated radio resources are provided to the mobile station in the PACKET DOWNLINK ASSIGNMENT message transmitted on the PCCCH.

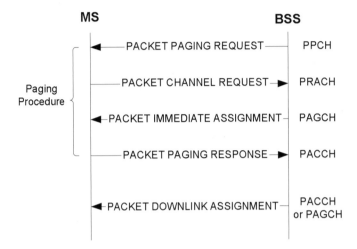

Figure 10.18 TBF Establishment Initiated by Network

10.5.2 Channel Access and Resource Allocation

The MAC is based on the time-slotted structure of the physical channel. There are two channel access methods:

- *Slotted Aloha method:* This random access method is applied only in PRACH where multiple mobile stations send request packets arbitrarily in the uplink direction at the beginning of a slot. In case the packet collides with a packet sent by another station, the station backs off for an arbitrary time and tries to transmit again.
- *Time Division Multiple Access (TDMA) method:* This contention-free channel access method is applied in all logical channels except PRACH. The packets are transmitted in predefined time slots allocated by the BSS. The standard defines three types of resource allocation mechanisms: Dynamic Allocation, Extended Dynamic Allocation, and Exclusive Allocation.

Using the USF mechanism described in the previous sections, the network can control the radio resource for each mobile station dynamically based on the demand of the mobile station (as indicated in the packet channel request) and the load of the network. USF is the 3-bit field included in the RLC/MAC header of the radio block (Figure 10.14) sent by the network in the downlink direction. This 3-bit field allows eight different USF values used for multiplexing in the uplink channel. On PCCCH, one USF value (USF = "111" or called USF = FREE) is used to indicate PRACH. The other USF values are used to reserve the uplink for different MSs.

10.5.2.1   Uplink Data Transmission

After TBF establishment, a TFI and assigned PDCHs are associated with the TBF. In addition, a USF is associated with an assigned PDCH. Up to eight TBFs can be multiplexed on the same PDCH. A mobile station listens and examines every RLC/MAC PDU on the assigned PDCHs (in the downlink). If the mobile station discovers its USF in the header of a RLC/MAC PDU, the following will occur. Depending on USF_GRANULARITY (assigned during the TBF establishment procedure) the MS will transmit starting at the next uplink block, either one or four consecutive radio blocks. Figure 10.19A shows an example of this procedure for the Dynamic Allocation medium access mode. The table in Figure 10.19B includes the TBFs at each MS. For TBF1 on MS1, USF GRANULARITY=0, resulting in one radio block transmission in the corresponding uplink block. In the case of TBF2 on MS2, USF GRANULARITY=1, resulting in four consecutive radio block transmissions.

Figure 10.19B shows an example of the USF procedure for the Extended Dynamic Allocation medium access mode where there are multiple PDCHs assigned to one connection. Notice that in this case if USF GRANULARITY=1, the four radio blocks will be transmitted on different PDCHs.

10.5.2.2   Downlink Data Transmission

For downlink transmissions which are governed only by the BSS, there is no need for the USF mechanism. The BSS will choose the appropriate packets for transmission from the TBF queues which reside at the BSS. This transmission will occur on the downlink PDCHs associated with the TBF.

10.5.3 Radio Link Control (RLC)

The RLC layer, which is located above the MAC layer, can provide reliability for MAC transmissions. The RLC can operate in one of the following three modes: transparent mode, acknowledged mode, and unacknowledged mode. The RLC in Iu mode includes all three modes while RLC in A/Gb mode includes only acknowledged mode and unacknowledged mode.

In transparent mode, the RLC has no functionality—that is, the packets pass through the RLC layer to the MAC layer without adding any protocol information.

In the acknowledged mode, the transfer of packets is controlled by a selective Automatic Repeat Request (ARQ) mechanism, in which the receiving station requests retransmission if an error occurs, coupled with the numbering of packets in one TBF. The sending station transmits packets within a window and the receiving station sends Packet Uplink Ack/Nack or Packet Downlink Ack/Nack messages. The ARQ mechanism implemented is Selective Repeat ARQ. In other words, every Ack message acknowledges all correctly received packets up to the indicated sequence number. The receiving station can selectively request erroneous packets for retransmission. The network may reallocate the network resources to accommodate the additional resources required for retransmissions. In EGPRS, acknowledged mode, the transmission of packets is also controlled by more advanced ARQ mechanisms that enable the retransmission of only part of the packet, reducing the retransmissions overhead.

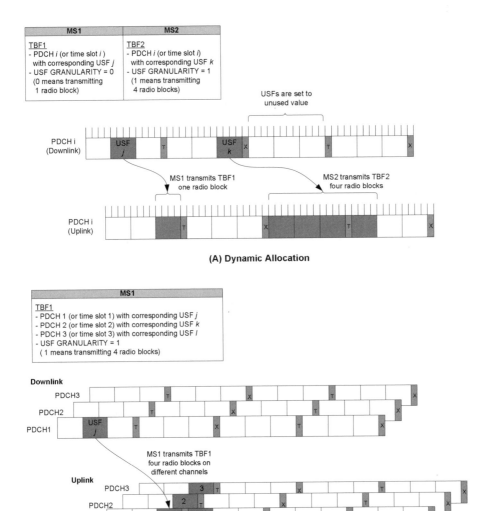

Figure 10.19 Examples of USF Mechanisms

In unacknowledged mode, the transmission of packets is controlled by the numbering of the packets within one TBF and does not include any retransmissions. The receiving station extracts user data from the received packets and attempts to preserve the user information length by replacing missing packets with dummy information bits.

10.5.4 MAC States

The MAC as applied to GERAN Iu mode can be in one of four states: MAC-Idle, MAC-Shared, MAC-DTM, and MAC-Dedicated.

In the MAC-Idle state there are no TBFs and the mobile station monitors relevant paging subchannels on the PCCCH. Once the upper layers have a packet for transmission and connect with the MAC, the MAC will establish a TBF on a shared channel, SBPSCH, and it transitions to MAC-Shared state. The RRC or RLC/MAC may also dictate to the MAC to establish a TBF on a dedicated channel, DBPSCH. In this case the MAC transits to a MAC-Dedicated state.

In the MAC-Shared state, the mobile station is allocated radio resources providing a TBF for a point-to-point, unidirectional transfer of upper layer packets on one or more shared channels (SBPSCHs). The MAC can transmit these packets in either RLC acknowledged mode or RLC unacknowledged mode. When all packets have been delivered and the acknowledgments have been received, then the MAC can release all TBFs on the downlink and uplink direction and return to the MAC-Idle state.

In the MAC-Dedicated state, the mobile station is allocated radio resources to transmit a TBF on one or more dedicated channels (DBPSCHs). The MAC can transmit these packets in RLC acknowledged mode, RLC unacknowledged mode, or transparent mode. After all packets have been delivered and acknowledgments have been received, the MAC can release all TBFs on the DBPSCHs and return to the MAC-Idle state.

In the MAC-Shared state, upper layers may require the transfer of an upper layer packet, which may trigger the establishment of a TBF on a dedicated channel, DBPSCH. In this case the MAC will transmit from its MAC-Shared state to the MAC-DTM state.

In the MAC-DTM state a mobile station has been allocated both shared and dedicated radio channels (one or more DBPSCHs and one or more SBPSCHs). Packets can be delivered in RLC acknowledged mode, RLC unacknowledged mode, or RLC transparent mode. When all packets that triggered the allocations of shared resources have been transmitted and consequent TBFs on the downlink and uplink SBPSCHs have been released, the MAC will transit to MAC-Dedicated state. Similarly, when all packets that triggered the allocation of dedicated channels have been transmitted and their consequent TBFs have been released to the MAC, the MAC will transit to MAC-Shared sate. Once all TBFs on both SBPSCHs and DBPSCHs have been released, the mobile station enters the MAC-Idle state.

When the MAC is in either MAC-Idle state or MAC-Shared state it continuously monitors the system information broadcasted in the PCCCH. The MAC on the mobile station may indicate to the RRC the availability of a new cell and a cell change. When a cell change is required either by the mobile station or by the network, the mobile station can operate in MAC-Idle state or MAC-Shared state in the current cell while getting system information on the new cell. However, the mobile station may suspend its TBF(s) and monitor the PCCCH in the current cell, in order to receive the necessary information on the new cell BCCH.

Once the switch to the new cell has been completed, the mobile station aborts any TBF in progress on both the downlink and uplink. The mobile station will reestablish its required TBF connections in the new cell.

10.6 Radio Resource Control (RRC) and Radio Resource (RR)

The Radio Resource Control (RRC) is responsible for radio resource management for Iu traffic, whereas the Radio Resource (RR) is responsible for managing the A/Gb traffic. The relationship between the two is presented in Figure 10.20.

The RRC and RR are responsible for allocating new dedicated basic physical subchannels as well as the intracell handovers of the dedicated basic physical subchannels.

The RRC broadcasts system information (such as location services in cells) to the mobile stations. The RRC also establishes, maintains, reconfigures, and releases connections between the mobile stations and the GERAN network. This includes cell reselection, admission control, and link establishment. This also includes selection of the connection parameters considering both control and applications needs. This allocation is then indicated to the mobile station. The RRC also handles mobility, which includes evaluation, decision, and execution of functions such as handover and moving from cell to cell. The RRC handover procedures control the mobility of the mobile station and include procedures to modify the channels allocated to the mobile station. Executing these functions, the RRC considers the requested QoS and ensures allocation of sufficient resources for the targeted QoS. In addition, the RRC performs monitoring, reporting, ciphering, and mediating between GERAN and the mobile stations.

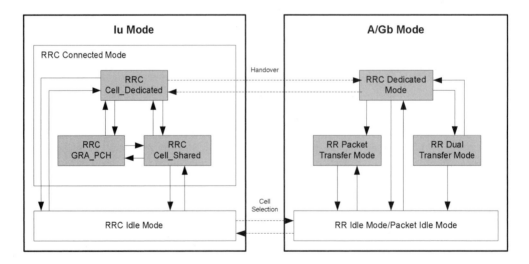

Figure 10.20 Relationship between RR and RRC

The RR maintains at least one PDCH that carries user data and all the necessary control signaling for initiating packet transfer whenever that signaling is not carried by the existing control channels. Other PDCHs, acting as slaves, are used for user data transfer and for dedicated signaling. The GPRS network is based on dynamically allocating capacity and hence does not require permanently allocated PDCHs. Capacity allocation is based on actual needs. A mobile station can be allocated permanent or temporary physical resources (i.e., PDCHs). In cases where some PDCHs are congested, the network may allocate more resources. The existence of PDCHs does not imply the existence of PCCCH. In this case, the GPRS stations can use the GSM control channel CCCH. The network can then assign resources on PDCHs for uplink transfer. After the transfer, the MS returns to CCCH. However, when PCCCH is allocated in the cell, all GPRS stations will use it. The network can allocate a PCCCH either as a result of the increased demand for packet data transfers or if there are enough available physical channels. If the network finds that the PCCCH capacity is not adequate, it can allocate additional PCCCH resources on one or several PDCHs.

The RRC has several modes of operation. When the mobile stations powers on and the Iu mode is selected, the MS enters the RRC-Idle Mode. In this mode the MS monitors control broadcasts messages. Once a connection is established and the Iu mode is entered, the RR-Idle mode changes to RRC-Connected mode. Such a connection can be made only after the upper layers have requested a connection and the network has responded by assigning communication resources. The RRC-Connected mode is characterized by three states: RRC-Cell_Shared, RRC-Cell_Dedicated, and RRC-GRA_PCH.

The RR supports both GPRS packet traffic and GSM traffic circuit switched traffic. The circuit switch part includes an Idle mode and Dedicated mode. The GPRS part includes Packet Idle mode and Packet Transfer mode. Stations belonging to Class A, supporting both GPRS and GSM circuit switching traffic, are in Dual Transfer mode.

10.6.1 RRC Modes of Operation

The RRC has several modes of operation: The RRC-Cell_Shared, RRC-Cell_Dedicated, and RRC-GRA_PCH.

In RRC-Cell_Shared state the MS executes a cell update procedure on cell changes, monitors the PBCCH control channel for system information messages, and monitors neighboring cells for neighbour cell measurements. The RRC will transit to RRC-Idle mode when the RRC connection is released or when the operation mode is changed to A/Gb mode. In the RRC-Cell_Shared state the MAC is responsible for allocating the shared physical subchannels.

The RRC-Cell_Dedicated state is assigned when a Dedicated Basic Physical Sub Channel (DBPSCH) is allocated to the MS. In this mode the MS is assigned one or more dedicated basic physical subchannels in the uplink and downlink. A station in this state performs measurement and reporting activities, listens to neighboring cells for measurements, and executes handover procedures of the dedicated basic physical subchannels on a cell change. Transition from RRC-Cell_Dedicated state to RRC-Idle mode occurs when the RRC connection is released. In the RRC-Cell_Dedicated state the RRC is responsible for allocating the physical dedicated subchannels.

The RRC-Cell_Dedicated state changes to RRC-Cell_Shared state when 1) all the dedicated basic physical subchannels are released and shared basic physical subchannels exist or 2) no shared basic physical subchannels exist.

The RRC-GRA_PCH state is assigned when GERAN orders the MS to move to RRC-GRA_PCH state via explicit signaling. In this transition the MS aborts any TBF in progress. In RRC-GRA_PCH state, no basic physical subchannel is allocated to the MS and no radio resource allocation tasks are executed. Hence, no uplink activity is possible. However, the MS continues to monitor the control channel; PCCCH and its location is known to GERAN. RRC will change from RRC-GRA_PCH state to RRC-Cell_Shared state due to changes in GRA update, cell update, or response to paging. Also in the RRC-GRA_PCH state, the MS may request a radio resource to answer to a paging message or to perform a GRA/Cell update procedure.

The RRC in RRC-Connected mode is responsible for allocating dedicated basic physical subchannels, which causes the MS to enter the RRC-Cell_Dedicated state. The MAC is responsible for allocating shared basic physical subchannels (SBPSCH). The MAC allocates the PDTCHs according to the QoS class of the radio link and the multi-slot capability of the MS. The RRC provides the MAC with QoS class and indication of the MS multi-slot capability.

10.6.2 RR Modes of Operation

The RR supports both GPRS packet switched traffic and GSM circuit switched traffic. The circuit switched part includes an idle mode and a dedicated mode as well as other modes related to voice services that are not discussed in this book. The GPRS part includes packet idle mode and packet transfer mode.

10.6.2.1   Packet Idle Mode

In packet idle mode, the MS listens to the PBCCH and to the paging subchannel for the paging messages. If PCCCH is not present in the cell, the mobile station listens to the BCCH and to the relevant paging subchannels. A station in the packet idle mode that belongs to GPRS class A may simultaneously enter the different RR service modes defined for GSM circuit switch services (e.g., circuit switched connections management, mobility management, and radio resource management). In this mode upper layers can require the transfer of packets that results in the establishment of TBFs and transition to packet transfer mode. This mode is not applicable to an MS supporting DTM with an active RR connection.

10.6.2.2   Dedicated Mode

An MS that supports DTM, meaning supporting both GPRS and other GSM services simultaneously, and has an active RR connection and has no allocated packet resources is in dedicated mode.

10.6.2.3   Packet Transfer Mode

In packet transfer mode, the mobile station is allocated radio resources for its TBFs on one or more physical channels on both the uplink and downlink. Delivery of the packets can be done in RLC acknowledged or RLC unacknowledged modes. Packet transfer mode is not applicable to a mobile station supporting DTM that has an active RR connection.

When selecting a new cell, the mobile station leaves the packet transfer mode and enters the packet idle mode. After moving to the new cell, it reads the system information and may change back to Packet Transfer Mode and resumes packet transmission operations. A GPRS class A station may simultaneously enter different RR service modes for GSM circuit switched services.

10.6.2.4   Dual Transfer Mode (DTM)

In dual transfer mode the mobile station supports an active RR connection and is allocated radio resources for its TBFs on one or more physical channels in the uplink and downlink directions. This mode is only applicable for a mobile station supporting GPRS or EGPRS and is a subset of class A mode of operation (i.e., supporting both GPRS and other GSM services simultaneously).

Packets can be transferred in RLC acknowledged or RLC unacknowledged mode. A station in DTM is carrying out all of the tasks of a station in the dedicated mode. In addition, the upper layers can require the release of all the packet resources, which triggers the transition to the dedicated mode, and the release of the RR resources, which triggers the transition to idle mode and packet idle mode.

When the station is handed over to a new cell, the RR leaves the DTM, enters the dedicated mode, may read the system information messages sent on the control channels, and then enters the DTM.

10.6.3 Transition between Modes

Figure 10.21 shows the four RR states for a class A station that does not support DTM. The four states can be regarded as a combination of two state machines with two RR states each: circuit switched part comprises idle mode and dedicated mode and the GPRS part comprises packet idle mode and packet transfer mode.

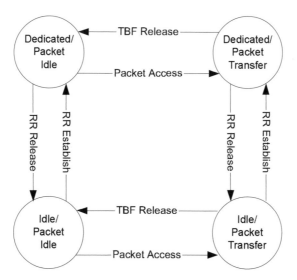

Figure 10.21 RR Modes for a Class A that Does Not Support DTM

Figure 10.22 shows the RR modes and transitions for a class A station that supports DTM and class B. Class B includes the following modes: packet idle mode, packet transfer mode, and dedicated mode.

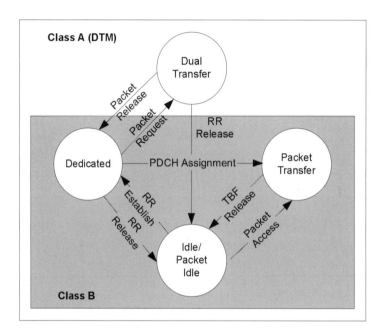

Figure 10.22 RR Modes and Transitions for Class B and Class A that Supports DTM

Figure 10.23 shows the RR modes and transitions for a Class C station. When it is attached to a GSM channel, there are two RR modes: idle mode and dedicated mode. When it is attached to a GPRS channel, there are two RR modes: packet idle mode and packet transfer mode.

10.7 QoS Support

The standard realizes that successful deployment of IP multimedia services requires support and management capability to provide different degrees of end-to-end QoS under various conditions. End-to-end QoS may be specified by Service Level Agreements. Hence, it requires consideration of available bandwidth, QoS classes, measurement, monitoring, and other tools. It requires that QoS support for IP multimedia sessions includes the following functions:

• Negotiate QoS for IP multimedia sessions as a whole as well as their individual components, when the session is established and during its execution. The assumption is that the IP multimedia applications are able to define their requirements, negotiate their capabilities, and identify and select the available media components, such as QoS.

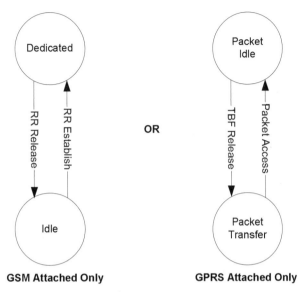

Figure 10.23 RR Modes and Transitions for Class C

- Allow either the user or the network to identify alternative destinations for IP multimedia sessions and their individual components. Based on this identification, these sessions can be redirected to these alternate destinations at various stages of the multimedia sessions.
- Support end-to-end QoS for voice that has quality at least as good as that provided by circuit-switched networks.
- Support roaming and negotiation between carriers for QoS.
- Deploy IP Policy Control for IP multimedia applications.
- Support more than one IP multimedia application in a session.
- Provide security services similar to those of circuit switched networks.
- Support for networking between the packet and circuit switched services.

GPRS specifies signaling that can enable support for various traffic streams with different characteristics such as constant bit rate or variable bit rate, connection or connectionless, point-to-point or point-to-multipoint, and bi-directional or unidirectional streams. These traffic streams can be established on a demand, reserved, or permanent basis.

The standards define several subscriber profiles that include information and parameters based on the contractual service agreement between the user and the carrier. The profile includes subscriber services and subscriber QoS profiles (e.g., service priority, reliability, delay, and throughput).

To provide end-to-end QoS, we need support from several components of the mobile system. The GPRS protocol architecture (control plane) is shown in Figure 10.24. We will focus on two QoS aspects: the GPRS bearer service QoS profile (Section 10.7.1) and radio bearer service QoS mechanisms (Section 10.7.2).

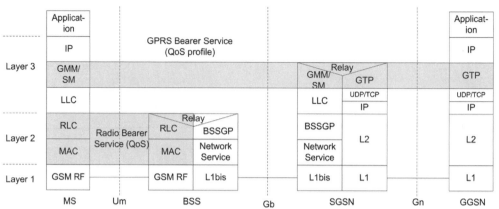

graphic
ok? (hyp
ation)

Figure 10.24 GPRS Protocol Architecture (Control Plane)

10.7.1 QoS Profile in GPRS Bearer Service

In order to provide QoS to an application, the network resources on each component of the mobile system (i.e., mobile station, Serving GPRS Support Node [SGSN], Gateway GPRS Support Node [GGSN]) need to be reserved. Resource reservation is carried out by the signaling mechanism that uses QoS profiles. A QoS profile is the set of QoS parameters that describes the application's characteristics and QoS requirements. A QoS profile consists of four parameters: service precedence (priority) parameter, reliability parameter, delay parameters, and throughput parameter. These parameters are described below.

10.7.1.1   Service Precedence or Priority Parameter

This includes the priority of supporting a specific traffic stream compared to other traffic streams. Higher priority streams will be served before lower priority streams. In case the network is congested, lower priority packets will be discarded before the higher priority packets. Three levels (classes) of priorities are applied (see Table 10.1).

Table 10.1 Service Precedence Classes

Precedence	Precedence Name
1	High Priority
2	Normal Priority
3	Low Priority

10.7.1.2   Reliability Parameter

This indicates the transmission characteristics required by the application in terms of loss probability, duplication of packets, mis-sequencing of packets, or corruption of packets. GPRS defines
three reliability classes that are described in Table 10.2.

Table 10.2 Reliability Classes

Reliability Class	Lost Packet Probability	Duplicate Packet Probability	Out of Sequence Packet Probability	Corrupt Packet Probability
1	10^{-9}	10^{-9}	10^{-9}	10^{-9}
2	10^{-4}	10^{-5}	10^{-5}	10^{-6}
3	10^{-2}	10^{-5}	10^{-5}	10^{-2}

10.7.1.3   Delay Parameter

This parameter defines the maximum values for the mean packet delay and the 95 percentile
delay that the GPRS network can provide. This delay can consist of the transmission delay from
the user to the network (uplink), or from the network to the user (downlink), and the delay in the
GPRS backbone. This delay does not include external networks or user higher layers. GPRS
defines five classes of delay values described in Table 10.3.

Table 10.3 Delay Classes

Delay Class	Delay (maximum values)			
	Packet Size: 128 octets		Packet Size: 1024 octets	
	Mean Transfer Delay (sec)	95 Percentile Delay (sec)	Mean Transfer Delay (sec)	95 Percentile Delay (sec)
1. (Predictive)	< 0.5	< 1.5	< 2	< 7
2. (Predictive)	< 5	< 25	< 15	< 75
3. (Predictive)	< 50	< 250	< 75	< 375
4. (Best Effort)	Unspecified			

10.7.1.4   Throughput Parameter

The throughput parameters include the maximum and mean bit rates. These terms can be negotiated between the user and the network before and during the communication session.

10.7.1.5   PDP Context Procedures

Each application's QoS requirements are mapped to a QoS profile. The mapping process is not defined in the standard. The QoS profile is included in the Packet Data Protocol (PDP) context. QoS profiles negotiation is managed by PDP context procedures (i.e., activation, modification, and deactivation). PDP context procedures are carried out by the Session Management (SM) protocol layer between MS and SGSN and by the GPRS Tunneling Protocol (GTP) layer between SGSN and GGSN as shown in Figure 10.25. Generally, PDP context procedures are used in PDP address (i.e., IP address) assignment and host configuration to a mobile station in addition to QoS profile signaling. Each application on a mobile station performs the PDP context procedures separately.

graphic ok? (hyp ation)

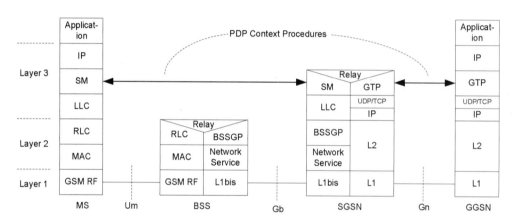

Figure 10.25 PDP Context Procedures

Figure 10.26 shows the PDP context activation procedure initiated by the MS. First, the MS sends SGSN an "Activate PDP Context Request" message that contains several information elements as well as the desired QoS profile of an application. The SGSN decides to accept or reject the request based on the available resources at the SGSN and subscribed QoS profile. If the SGSN accepts the request, it will send GGSN a "Create PDP Context Request" message containing the QoS profile. After receiving the "Create PDP Context Request" message, the GGSN decides to accept or reject the request based on the available resources at the GGSN and the mobile station's subscription information. If the GGSN accepts the request, GGSN sends the SGSN a "Create PDP Context Response" message including the negotiated QoS profile. Finally, SGSN includes the negotiated QoS profile in an "Activate PDP Context Accept" message and sends it to the MS.

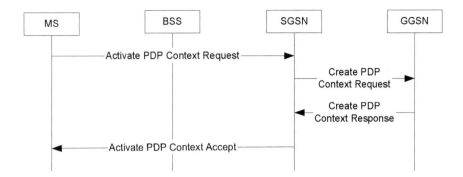

Figure 10.26 PDP Context Activation Procedure Initiated by MS

There are cases where packets originated from the external network are received by the GGSN and are destined to a mobile station. In these cases, the GGSN will initiate the PDP context activation procedure shown in Figure 10.27. In Step 2 the GGSN acquires the routing information to the MS from the Home Location Register (HLR). After receiving this routing information from the HLR, in Step 3 the GGSN sends a "PDU Notification Request" message to the SGSN and then the SGSN replies back with a "PDU Notification Response" to confirm that the SGSN will initiate a "Request PDP Context Activation" message. The SGSN sends the MS the "Request PDP Context Activation" message. In Step 5, the MS continues with a similar process as in PDP context activation initiated by the MS case (shown in Figure 10.26).

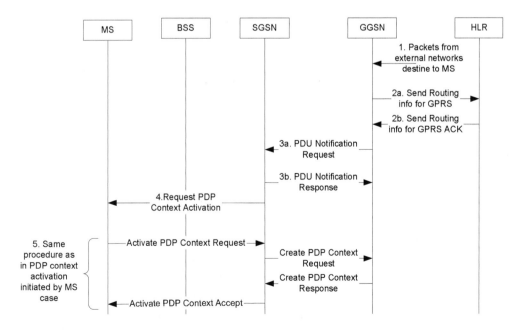

Figure 10.27 PDP Context Activation Procedure Initiated by GGSN

10.7.2 QoS Mechanisms in the Radio Bearer Service

In this subsection we will focus on the radio interface QoS mechanisms that enable quantitative QoS services: classification and packet scheduling. Figure 10.28 shows a simplified diagram of these QoS mechanisms.

First, an application performs the QoS negotiation with the SGSN as described in the previous subsection. When data packets arrive from the upper layer to the RLC/MAC, the TBF establishment process will be activated. During the TBF establishment process, the QoS profile will be mapped to RLC/MAC parameters (i.e., radio priority) included in the PACKET CHANNEL REQUEST. The mapping process is not defined in the standard. After the TBF is established, the TBF QoS parameters are used at the MS to classify the packets and at the BSS to make packet scheduling decisions. In the remaining part of this section we provide more details on the classification and packet scheduling mechanisms.

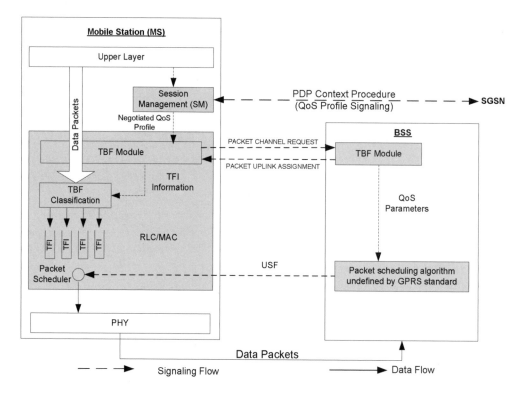

Figure 10.28 Simplified GPRS QoS Architecture on the Radio Interface

10.7.2.1   Classification

All packets that belong to a specific TBF are tagged with the corresponding TFI. These tags along with the direction (i.e., uplink, downlink) enable packet classification. Therefore, GPRS uses per-flow classification which supports per-flow QoS services. Notice that due to the temporary nature of the TBF, an application may have the TBF established and terminated several times during the course of the service. Therefore, TFI will be different for each TBF connection.

10.7.2.2   Packet Scheduling

As described in the Section 10.5.2, the channel access is based on TDMA. The time slot assignment is based on the request-grant process. The current bandwidth demand in terms of the number of RLC/MAC blocks waiting in queue is included in the PACKET CHANNEL REQUEST message sent during the TBF establishment process. Packet scheduling uses bandwidth demand information as well as TBF QoS parameters to decide when packets that belong to a specific TBF are allowed to transmit. Therefore, the packet scheduling algorithm can dynamically allocate the bandwidth based on an application's bandwidth demand and QoS requirements. The dynamic bandwidth allocation decisions performed at the BSS are communicated to the mobile stations using the USF process (see Section 10.5.2). It is important to mention that the packet scheduling algorithms are not defined by the standard. These algorithms should be defined and implemented by the GPRS network designers and carriers.

In summary, the GPRS system combined with proper packet classification and packet scheduling algorithms can enable per-flow quantitative QoS services.

CHAPTER 11

UMTS

11.1 Introduction

Universal Mobile Telecommunications System (UMTS) is a 3G technology that provides high-speed connection of up to 2 Mbps to support a wide variety of services, including multimedia services with different QoS requirements. As shown in Figure 11.1, UMTS evolved from 2G GSM technology and 2.5G GPRS. The effective bandwidth provided is up to 144 kbps for vehicular users, 384 kbps for mobile users, and up to 2 Mbps for static users. The recent UMTS Rel-5 added to previous releases (Rel-99 and Rel-4) the High-Speed Download Packet Access (HSDPA) technology. HSDPA adds an option of higher speeds of up to 10 Mbps required for multimedia applications.

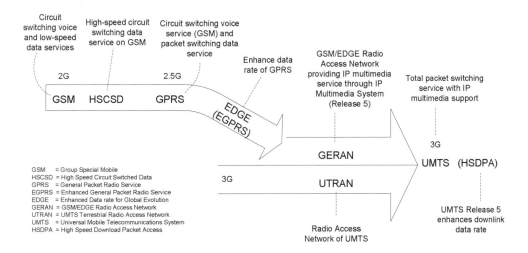

Figure 11.1 UMTS Evolution

UMTS, which is also referred to as WCDMA (Wideband CDMA or Wideband Code Division Multiple Access), is part of IMT-2000 (International Mobile Telecommunication–2000) effort, which is designed to provide high-speed communication with high-quality multimedia services and global roaming support. The IMT-2000 effort is led by the ITU (International Telecommunications Union), formerly the CCITT (Consultative Committee for International Telephony and Telegraphy), is an international organization that sets communication standards and consists of more than 150 member countries.

The IMT-2000 effort strives to create a family of compatible standards that can be used worldwide for all mobile applications. This family is expected to support both circuit and packet switched applications and provides QoS support for multimedia traffic. This effort considered several proposals for both terrestrial and satellite communication, from which WCDMA and CDMA 2000 were selected for terrestrial communication.

UMTS has been introduced and evolved in several releases. Starting with Release 99 UMTS integrated GSM and GPRS such that previous investments in those networks can be preserved. UMTS Terrestrial Radio Access Network (UTRAN) is UMTS's radio access network that provides the communication path between the mobile station and the 3G core network through the Iu interface. UMTS shares the same core network as GPRS except the UTRAN radio interface, which is newly designed (see Figure 11.2). Therefore, GPRS and UMTS are fully harmonized.

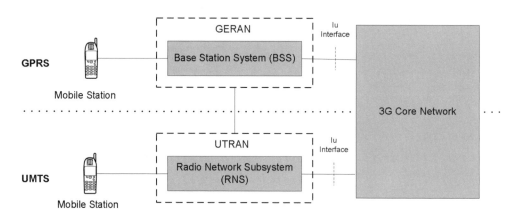

Figure 11.2 GPRS and UMTS

UTRAN supports two modes of operation: 1) Frequency Division Duplex (FDD), where uplink transmissions (from the mobile stations to UTRAN) and downlink transmissions (from UTRAN to the mobile stations) operate simultaneously using two separate frequency bands; and 2) Time Division Duplex (TDD), where uplink and downlink transmissions operate interchangeably in different time periods on the same frequency band.

In UTRAN FDD mode, UTRAN deploys Wideband Code Division Multiple Access (WCDMA), whereas in UTRAN TDD mode, UTRAN deploys Time Division–Code Division

Multiple Access (TD-CDMA). Figure 11.3 shows UMTS coverage scenarios. FDD mode, which is geared toward wide area coverage (i.e., public macro and micro cell), supports high user mobility and data rates of up to 384 kbps. TDD, which is geared toward smaller geographical areas (i.e., public micro and pico cells), supports data rates of up to 2 Mbps and slow user mobility (i.e., cordless phones). In order to operate in both wide and local areas, a mobile station or user equipment (UE) should support both TDD and FDD modes.

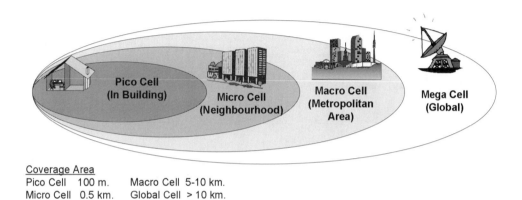

Coverage Area
Pico Cell   100 m.   Macro Cell 5-10 km.
Micro Cell   0.5 km.   Global Cell > 10 km.

Figure 11.3 UMTS Geographical Coverage

In this book we focus on UMTS Rel-5, which is presented in the following documents:

- TS 23.002 3rd Generation Partnership Project; Technical Specification Group Services and Systems Aspects; Network Architecture (Release 5)
- TS 25.301 3rd Generation Partnership Project; Technical Specification Group Radio Access Network; Radio Interface Protocol Architecture (Release 5)
- TS 25.401 3rd Generation Partnership Project; Technical Specification Group Radio Access Network; UTRAN Overall Description (Release 5)
- TS 25.302 3rd Generation Partnership Project; Technical Specification Group Radio Access Network; Services provided by the physical layer (Release 5)
- TS 25.211 3rd Generation Partnership Project; Technical Specification Group Radio Access Network; Physical channels and mapping of transport channels onto physical channels (FDD) (Release 5)
- TS 25.321 3rd Generation Partnership Project; Technical Specification Group Radio Access Network; MAC Protocol Specification (Release 5)
- TS 25.322 3rd Generation Partnership Project; Technical Specification Group Radio Access Network; Radio Link Control (RLC) Protocol Specification (Release 5)
- TS 25.323 3rd Generation Partnership Project; Technical Specification Group Radio Access Network; Packet Data Convergence Protocol (PDCP) Specification (Release 5)
- TS 25.331 3rd Generation Partnership Project; Technical Specification Group Radio Access Network; Radio Resource Control (RRC); Protocol Specification (Release 5)

• TS 25.308 3rd Generation Partnership Project; Technical Specification Group Radio Access Network; High-Speed Downlink Packet Access (HSDPA); Overall Description; Stage 2 (Release 5)

11.2 UMTS Architecture

UMTS Rel-5 provides the ultimate evolution of voice and data convergence. As shown in Figure 11.4, similar to GPRS architecture, UMTS consists of three main entities: Mobile Station (MS), UTRAN, and Core Network (CN). A mobile station or user equipment (UE) communicates with Node B which controls the radio channel within its coverage area or cell. Multiple Node Bs are controlled by the Radio Network Controller (RNC). UMTS's Node B is equivalent to GPRS's Base Transceiver Station (BTS), while RNC is equivalent to GPRS's Base Station Controller (BSC). RNC and associated Node B form the radio network subsystem (RNS). The RNCs are interconnected to each other through the Iur interface. RNC connects to the core network through Iu interface which supports both voice and data services.

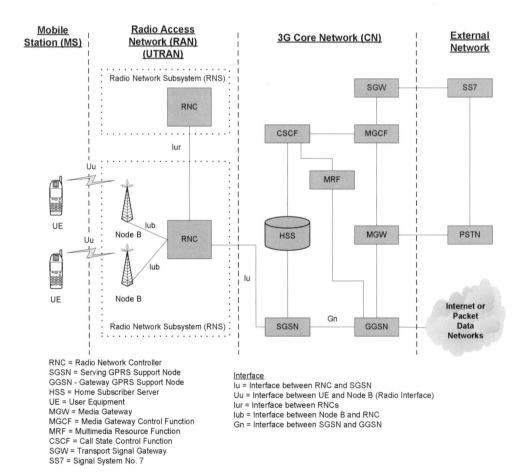

RNC = Radio Network Controller
SGSN = Serving GPRS Support Node
GGSN - Gateway GPRS Support Node
HSS = Home Subscriber Server
UE = User Equipment
MGW = Media Gateway
MGCF = Media Gateway Control Function
MRF = Multimedia Resource Function
CSCF = Call State Control Function
SGW = Transport Signal Gateway
SS7 = Signal System No. 7

Interface
Iu = Interface between RNC and SGSN
Uu = Interface between UE and Node B (Radio Interface)
Iur = Interface between RNCs
Iub = Interface between Node B and RNC
Gn = Interface between SGSN and GGSN

Figure 11.4 UMTS Architecture

UTRAN's protocol stack presented in Figure 11.5 consists of the Physical Layer (Layer 1), Data Link layer (Layer 2), and Network layer (Layer 3). The Physical Layer (PHY) includes the radio. The Data Link includes the Media Access Control (MAC), the Radio Link Control (RLC), the Packet Data Convergence Protocol (PDCP), and the Broadcast/Multicast Control (BMC). The Network Layer includes the Radio Resource Control (RRC). The RRC and RLC are divided into Control and User planes. PDCP and BMC exist in the User plane only. The PDCP and RLC in Rel-5 are unchanged from the Rel-99 and Rel-4 architecture, whereas the MAC is changed to include a new module that supports HSDPA. Layers 1, 2, and 3 are part of UMTS's Access Stratum (AS). The higher layer that is part of UMTS's Non-Access Stratum (NAS) includes Mobility Management (MM), Call Control (CC), and Session Management (SM).

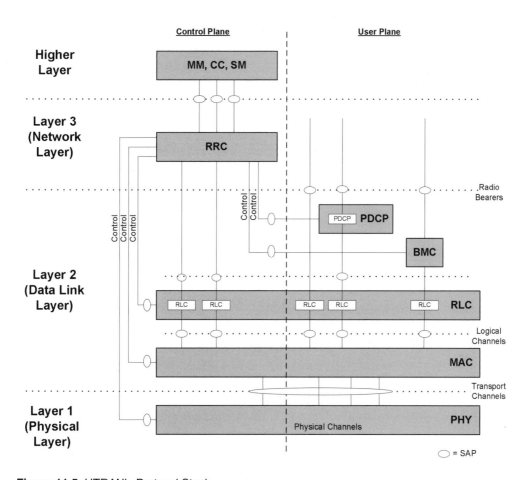

Figure 11.5 UTRAN's Protocol Stack

The transport channels which indicate how the data are transferred over the radio interface are defined by the service access point (SAP) between the MAC and PHY layers. The logical channels which indicate the contents or types of data transferred are defined by the SAP between the RLC and MAC layers. Therefore, one of the MAC's functions is to map the logical channels to the transport channels. The services provided by layer 2 are considered to be the radio bearer.

The Physical Layer, which is described in detail in the next section, provides transport services to the MAC through the transport channels. The physical layer transport channels are mapped to the physical channels by defining a number of radio mechanisms (i.e., modulation, channel coding, radio matching, multiplexing, interleaving). The MAC layer contains several MAC entities (i.e., MAC-d, MAC-c/sh) which will be described in Section 11.4. The Radio Link Control (RLC), the Packet Data Convergence Protocol (PDCP), the Broadcast/Multicast Control (BMC), and the Radio Resource Control (RRC) are introduced in Section 11.5.

11.3 Physical Layer

The PHY provides data transfer services to the MAC and higher layers. The PHY is responsible for 1) various handover functions, 2) error detection and report to higher layers, 3) multiplexing of transport channels, 4) mapping of transport channels to physical channels, 5) power control, 6) synchronization in TDD mode, and 7) other radio responsibilities associated with transmitting and receiving signals over the wireless media.

The PHY's UMTS Terrestrial Radio Access Network (UTRAN) supports the following two modes: FDD (Frequency Division Duplex) mode and TDD (Time Division Duplex) mode. In the FDD mode, the PHY radio works in two frequency bands (each band has 5 MHz channel bandwidth), one band for uplink transmission and the other for downlink transmission. In TDD mode, the same frequency band (with a 5 MHz channel bandwidth) is used for both uplink and downlink transmissions. The uplink and downlink operate reciprocally in different time periods. Figures 11.6 and 11.7 illustrate the 3GPP (3G Partnership Project) frequency spectrum for the FDD and TDD modes, respectively. However, some countries may define their own frequency bands for 3G.

The FDD mode allows symmetric data transmissions on the uplink and downlink. On the other hand, the TDD mode has more flexibility in assigning asymmetric uplink and downlink transmissions. Therefore, TDD is suitable to support asymmetric services, for example where most of the traffic is on the downlink (from the backbone to the mobile users). However, since TDD requires time synchronization, it cannot be deployed in wide coverage areas (i.e., mega cell) where the large variations in propagation delays do not allow tight synchronization between the nodes.

For channel access, the PHY deploys Wideband Direct Sequence Code Division Multiple Access (WCDMA), in which the information bits spread over 5 MHz channel bandwidth using a number of chip signals. The 3.84 Mcps chip rate is used in both FDD and TDD modes. In addition, TDD has an option of 1.28 Mcps chip rate, which spreads the information bits over 1.6 MHz channel bandwidth. The channel access of the latter case is denoted as narrowband

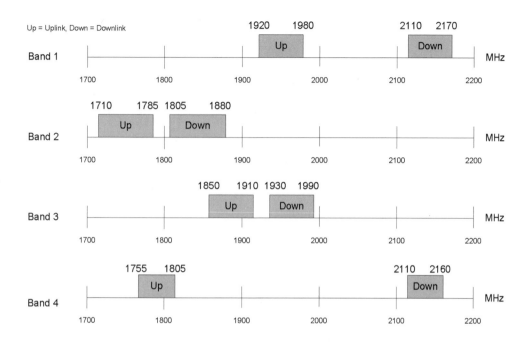

Figure 11.6 UMTS FDD Frequency Bands

CDMA. The key concept of code division multiple access is that multiple users can transmit data simultaneously on the same frequency band using different codes. Figure 11.8 illustrates the channel access of the FDD and TDD modes.

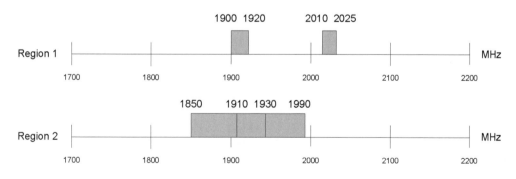

Figure 11.7 UMTS TDD Frequency Bands

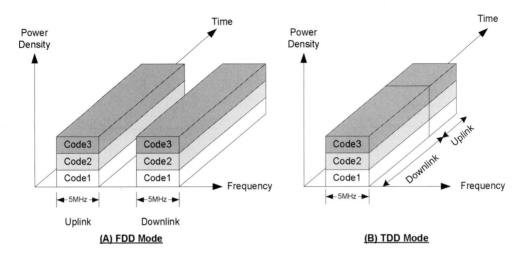

Figure 11.8 Channel Access

The code division multiple access involves two types of codes: scrambling code and channelization code. Figure 11.9 illustrates the simplified diagram of the coding process. The data from a physical channel are coded with channelization code and scrambling code and then transmitted. At the receiver end, the same codes (both channelization and scrambling) are used to decode the signal.

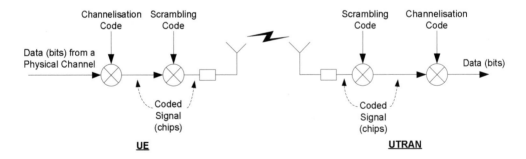

Figure 11.9 Channelization and Scrambling Process

The scrambling code is based on the complex-valued Gold Code sometimes called Pseudo Noise (PN) code. The scrambling code is used to separate between transmissions from different nodes. For example, when multiple transmissions arrive simultaneously on the uplink, Node B can retrieve the data signal from a specific mobile station by applying the same scrambling code of the mobile station to the incoming signal. On the downlink, where only Node Bs transmit, the scrambling code can also be applied to separate the data signals from different cells. For the uplink, there is a large number of scrambling codes (i.e., in the order of several millions) to be assigned to the mobile stations. For the downlink, the standard limits the number of codes to only 512. One of the reasons for this limited number of codes is to reduce the time it takes a new station to join a cell. When a new station joins a cell, the station has no knowledge of what scrambling code is used by Node B. The station has to scan the channel to find out this code. Obviously, searching only 512 codes, as opposed to millions of codes, expedites this search process.

The channelization code is based on Orthogonal Variable Spreading Factor (OVSF) codes, sometimes called Walsh codes. The channelization code is used to separate the sources of different physical channels on a mobile station. The physical channels will be described in a later section. Different physical channels on a mobile station deploy different OVSF codes. It is possible to have the same OVSF code assigned to physical channels on different mobile stations. An OVSF code that contains a sequence of chips spreads the data signal (in bits) to a higher bandwidth data signal (in chips). OVSF codes are organized in a tree structure as shown in Figure 11.10.

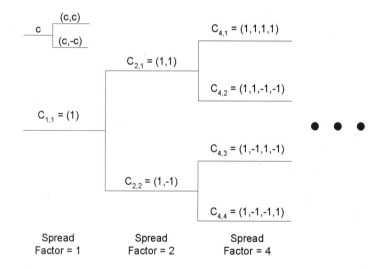

Figure 11.10 OVSF Code Tree

The OVSF code tree contains the hierarchical structure of codes where each level is referred to the spread factor (SF). For example, $SF = 4$ contains four OVSF codes with the length of four chips each. Codes within the same spread factor are orthogonal to each other. The code assignment for the uplink is managed by each mobile station, and by the RNC on the downlink. There are some restrictions for the code assignment. A code can be assigned to a physical channel if there is no other physical channel that is allocated this code or codes in the underlying branch. For example (see Figure 11.10), code $C_{2,1}$ can be assigned if code $C_{2,1}$ and codes in the underlying branch (i.e., $C_{4,1}$, $C_{4,2}$, $C_{8,1}$, $C_{8,2}$, $C_{8,3}$, $C_{8,4}$ and so on) are not assigned to any physical channel. Furthermore, smaller spread factor codes on the path to the root cannot be used (i.e., $C_{1,1}$).

The level of signal spreading is determined by the spread factor. For example, a spread factor equal to 256 allows 256 OVSF codes (256 chip length each code). The higher the spread factor is, the lower the symbol data rate (or bit rate) but the higher the number of physical channels that can transmit simultaneously. The possible spread factor numbers are 1, 2, 4, 8, 16, 32, and so on.

In FDD mode, the spread factors are from 256 to 4 on the uplink and from 512 to 4 on the downlink. With 3.84 Mcps chip rate, in FDD, the information rate varies from 15 ksymbols/s (3.84 M/256) to 960 ksymbols/s (3.84 M/4) on the uplink, and from 7.5 ksymbols/s (3.84 M/512) to 960 ksymbols/s on the downlink.

In TDD mode, the spread factors range from 16 to 1 on both the uplink and downlink. In TDD, the information rate varies from 240 ksymbols/s (3.84M/16) to 3.84 Msymbols/s.

The assigned data rate or bandwidth to a connection is controlled by the spread factor. The spread factor assignment can be fixed or dynamically adjusted based on the bandwidth demand of the connection. The lower the spread factor is, the higher the achievable data rate of the connection. Another approach to provide higher data rates is to assign multiple physical channels to a connection.

An important issue we need to consider in WCDMA is the power control (especially on the uplink). After the incoming signal is appropriately decoded (i.e., scrambling and channelization decoding), the desired signal is eminent from other signals (which are considered as noise interference). If the power density of the undesired signals (other channels) is too high, the decoded signal will contain a high level of noise interference and eventually compromise the desired signal detection. Therefore, without power control, an overpowered station could interfere or block the signals from other stations. Figure 11.11 illustrates power control in WCDMA.

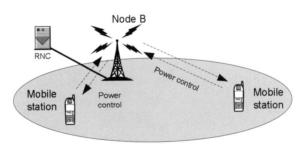

Figure 11.11 Power Control in UMTS

The standard defines two types of power control: open-loop power control and fast closed-loop power control. In the open-loop power control mechanism, a mobile station estimates the signal strength from the downlink beacon signal and sets its uplink strength power. This power control is inaccurate and often used in initial power setting. In the fast closed-loop power control mechanism, Node B measures the signal-to-interference ratio (SIR) from the mobile station. Then, Node B reports this SIR measurement to the RNC, which in turn determines the appropriate value of the mobile station's signal power. The RNC sends the Transmit Power Control (TPC) command (included in the physical control channel) to the mobile station. The mobile station readjusts the transmission power according to the received TPC command.

UTRAN deploys the QPSK modulation scheme. 8-PSK is used for 1.28 Mcps TDD option, whereas 16-QAM is used in HS-DSCH.

11.3.1 Transport and Physical Channels

One of the functions of the physical layer is to perform the mapping process between the transport and physical channels. As shown in Figure 11.12, the MAC packets on a transport channel are called transport blocks. A number of transport blocks transmitted on the same transport channel are called a Transport Block Set. A Transport Block Set size is defined as the number of bits in a Transport Block Set. As shown in Figure 11.13, the interarrival time between consecutive Transport Block Sets is defined as the Transmission Time Interval (TTI).

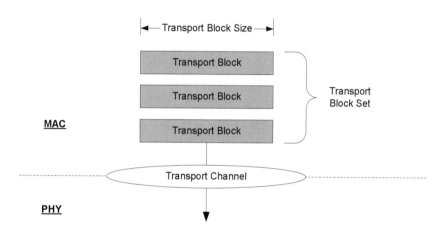

Figure 11.12 Transport Block

The Transport Block Size, the Transport Block Set Size, and the Transmission Time Interval are included in the Transport Format. The Transport Format contains two parts: the dynamic and semi-static parts. The dynamic part contains the Transport Block Size and the Transport Block Set Size. The semi-static part contains the Transmission Time Interval, the error protection scheme (channel code, coding rate, rate matching), and the CRC size. There are a number of Transport Format combinations called Transport Format Combination Set (TFCS). For example, a TFCS is shown in Tables 11.1 and 11.2 for dynamic and semi-static parts, respectively. The TFCS is assigned by the Radio Resource Control (RRC). The MAC selects an appropriate Transport Format Combination (TFC) from the TFCS to support the connection's traffic characteristics and data rate. The MAC uses dynamic attributes for the TFC and TTI, while the PHY uses semi-static attributes for the TFC.

TTI = Transmission Time Interval

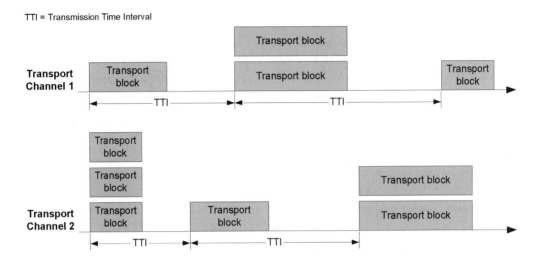

Figure 11.13 Transport Time Interval

Table 11.1 Transport Format Combination Set Dynamic Part

	Transport Channel 1		**Transport Channel 2**		**Transport Channel 3**	
	Transport block size (bits)	**Transport block set size (bits)**	**Transport block size (bits)**	**Transport block set size (bits)**	**Transport block size (bits)**	**Transport block set size (bits)**
Combination 1	20	20	100	200	120	240
Combination 2	40	40	160	160	320	320
Combination 3	320	1280	40	40	320	1280

Table 11.2 Transport Format Combination Set Semi-Static Part

	Transport Time Interval (ms)	**Type of Error Protection Code**	**Static Rate Matching Parameter**
Transport Channel 1	10	Convolution code	1
Transport Channel 2	20	Turbo code	1
Transport Channel 3	30	Turbo code	2

Figure 11.14 shows the physical layer mapping process. In a certain period of time, transport blocks arrive from the MAC along with Transport Format Indicator (TFI) which indicates the transport format of the transport blocks. The transport blocks from different transport channels are multiplexed, channel coded (i.e., convolution code, turbo code), split, and passed to the physical data channels. The TFIs are also multiplexed onto the Transport Format Combination Indicator (TFCI) and then passed to a physical control channel. There can be one or multiple physical data channels but there is only one physical control channel. At the receiver end, after descrambling and channelization decoding, the receiver passes the incoming packets to the decoding, demultiplexing, and channel joining module which uses the information within the TFCI. The transport blocks are routed to the appropriated transport channels to the higher layer.

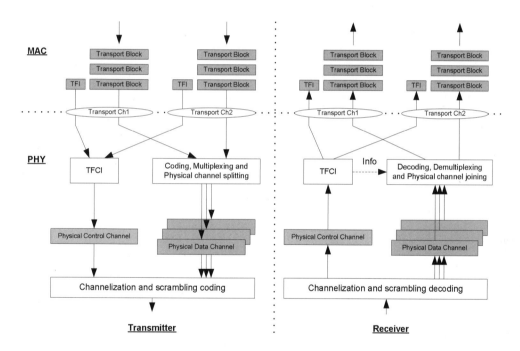

Figure 11.14 Mapping Process at the Physical Layer

11.3.1.1 Transport channels

The transport channels (see Figure 11.15) that transmit the data through the radio interface can be categorized as either dedicated transport channels or common transport channels. Dedicated transport channels are identified by the physical channel—that is, 1) code and frequency in FDD mode, or 2) code and time slot in TDD mode. Common transport channels are identified (if required) by an in-band identification.

Since network resources are reserved for the dedicated transport channels, these channels are suitable for real-time applications. Dedicated Channel (DCH) is a channel dedicated to one mobile station and is used in either uplink or downlink direction. DCH may contain user or control data from the upper layer.

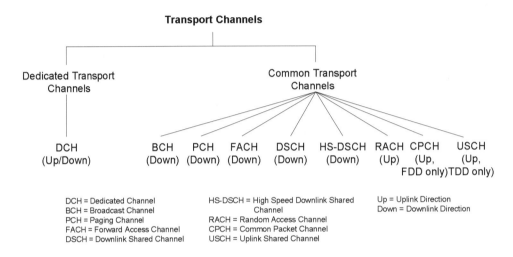

Figure 11.15 Transport Channels

Common transport channels include:

- Broadcast Channel (BCH) is a downlink channel used to broadcast system information to the entire cell such as the available access code and access time slots used for the random access mechanism. BCH's transmission power should be sufficient to reach all mobile stations within a cell.
- Paging Channel (PCH) is a downlink channel used to broadcast control information to the entire cell in order for the mobile station to execute its sleep mode procedures.
- Forward Access Channel (FACH) is a common downlink channel used to transmit relatively small amounts of data (i.e., SMS message) or control messages. FACH may contain messages destined to several mobile stations. Therefore, in-band identification is required. FACH provides low bit rate and does not perform fast closed-loop power control.
- Downlink Shared Channel (DSCH) is a downlink channel shared by several mobile stations. It is used for transmission of dedicated user or control data. DSCH provides variable bit rate and is required to perform fast closed-loop power control.
- High-Speed Downlink Shared Channel (HS-DSCH) is a downlink channel shared between mobile stations by allocation of individual codes, from a common pool of codes assigned for the channel. It is similar to DSCH but provides higher data rates.

- Random Access Channel (RACH) is a contention based uplink channel used for transmission of relatively small amounts of data (e.g., for initial access or non-real-time control or user data). RACH does not perform fast closed-loop power control. Due to unpredictable access delays of the contention based scheme, RACH is not suitable for real-time control and user data.
- Common Packet Channel (CPCH) is a contention based uplink channel shared by the mobile stations used for transmission of bursty data traffic. CPCH provides variable bit rate and is required to perform fast closed-loop power control. (*Note: This channel is used in FDD mode only.*)
- Uplink Shared Channel (USCH) is an uplink channel shared by several mobile stations. It is used for dedicated control or user data. (*Note: This channel is used in TDD mode only.*)

In summary, traffic can be transmitted through:

- Uplink

  - RACH: contention based, small amounts of data, no power control
  - CPCH (FDD mode only): contention based, more data than RACH, power control required
  - USCH (TDD mode only)
  - DCH: contention free, dedicated bandwidth, large amounts of data, power control required and QoS support

- Downlink

  - FACH: small amounts of data, shared with multiple users
  - DSCH: more data than FACH, shared with multiple users
  - DCH: dedicated bandwidth to a user and QoS support

11.3.1.2   Physical Channels

The physical channels are defined in the physical layer by a specific channel frequency, channelization code, and time duration. The standard defines the structure of the physical channel in radio frames and time slots. A radio frame has the duration of 38400 chips, which is equivalent to 10 ms at 3.84 Mcps chip rate. A radio frame consists of 15 time slots and the duration of each time slot corresponds to 2560 chips. Each time slot is numbered from 0 to 14 (see Figure 11.16). This frame structure is applied to both TDD and FDD modes. A physical channel can be transmitted on one or multiple time slots (consecutively or nonconsecutively). Multiple physical channels with different channelization codes can coexist in the same time slot. Some physical channels may have different structures.

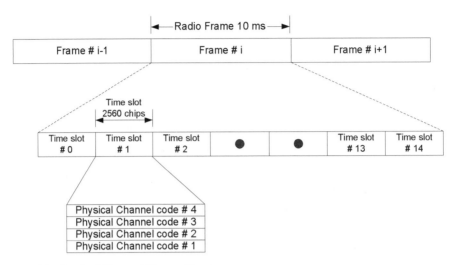

Figure 11.16 Physical Channel Frame Structure

Figure 11.17 shows a frame structure of the UTRAN TDD mode with several switching point configurations.

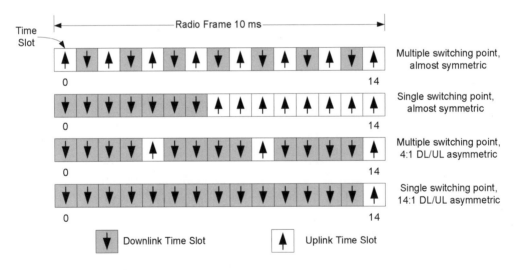

Figure 11.17 UTRAN TDD Frame Structure

The standard defines several physical channels to support the transport channels. Some physical channels are used for Layer 1 control exchange between UE and UTRAN and are not visible to the upper layer.

The FDD physical channels include:

- Dedicated Physical Data Channel (DPDCH)
- Dedicated Physical Control Channel (DPCCH)
- Physical Random Access Channel (PRACH)
- Physical Common Packet Channel (PCPCH)
- Common Pilot Channel (CPICH)
- Primary Common Control Physical Channel (P-CCPCH)
- Secondary Common Control Physical Channel (S-CCPCH)
- Synchronization Channel (SCH)
- Physical Downlink Shared Channel (PDSCH)
- Acquisition Indicator Channel (AICH)
- Access Preamble Acquisition Indicator Channel (AP-AICH)
- Paging Indicator Channel (PICH)
- CPCH Status Indicator Channel (CSICH)
- Collision-Detection/Channel-Assignment Indicator Channel (CD/CA-ICH)
- High-Speed Physical Downlink Shared Channel (HS-PDSCH)
- HS-DSCH-related Shared Control Channel (HS-SCCH)

The TDD physical channels include:

- Dedicated Physical Channel (DPCH)
- Primary Common Control Channel Physical Channel (P-CCPCH)
- Secondary Common Control Physical Channel (S-CCPCH)
- Physical Random Access Channel (PRACH)
- Physical Uplink Shared Channel (PUSCH)
- Physical Downlink Shared Channel (PDSCH)
- Paging Indicator Channel (PICH)
- Synchronization Channel (SCH)

Figure 11.18 shows the mapping of the transport channels to the physical channels.

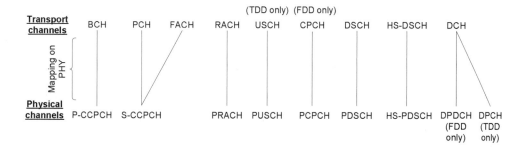

Figure 11.18 Transport Channels Mapped onto Physical Channels

In this chapter we cover the following mechanisms: dedicated data transmissions and random access uplink data transmissions.

11.3.2 Dedicated Data Transmission

DCH which supports dedicated data transmissions (uplink and downlink) is mapped to DPDCH and DPCCH as shown in Figure 11.19.

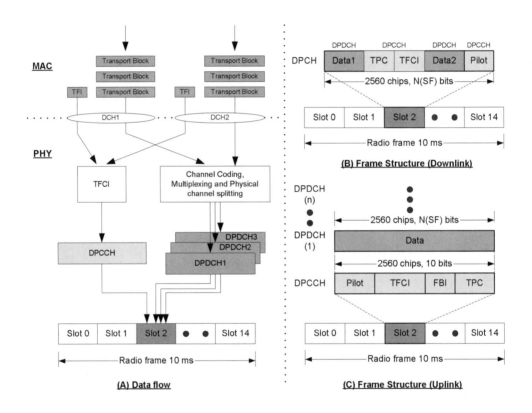

Figure 11.19 DCH Mapping Process and Physical Channel Frame Structure

11.3.2.1   Uplink

DPCCH carries control information generated at the physical layer. A pilot bit is used for channel estimation in coherent detection. Transport Format Combination Indicator (TFCI) notifies the receiver about the format combination of transport channels mapped to DPDCHs. Transmit Power Control (TPC) is used for fast close-loop power control.

DPDCH contains user data (transport block) from the upper layer. FBI contains feedback information. DPCCH uses a fixed channelization code with SF = 256. Therefore, a DPCCH frame contains 10 bits (2560/256).

DPDCH carries the user data (transport blocks) from DCHs. There are up to six DPDCHs in a time slot. If only one DPDCH exists, the DPDCH uses a channelization code with SF ranging from 256 to 4. If multiple DPDCHs exist, DPDCHs use a channelization code with SF = 4. It is flexible to vary the data rate of dedicated data transmissions using varying channelization code (from frame to frame) and a number of concurrent DPDCHs. On the uplink, each time slot contains one or more DPDCHs (up to six) as well as a DPCCH.

11.3.2.2   Downlink

DPCH, the downlink dedicated physical channel, contains the multiplexing of DPDCHs and DPCCHs. DPCH uses a channelization code with SF ranging from 512 to 4.

11.3.3 Random Access Uplink Data Transmission

RACH and CPCH support random access data transmission (uplink only). CPCH is used only in FDD mode. RACH is mapped to PRACH whereas CPCH is mapped to PCPCH. Figures 11.20 and 11.21 show the mapping process of RACH and CPCH, respectively. Each mobile station has only one RACH and one CPCH. In RACH, the control part contains TFCI that provides the decoding information used for the data part of the PRACH. In CPCH, the control part also contains TPC (for power control), pilot, and FBI in addition to TFCI. Both control and data part use different channelization codes.

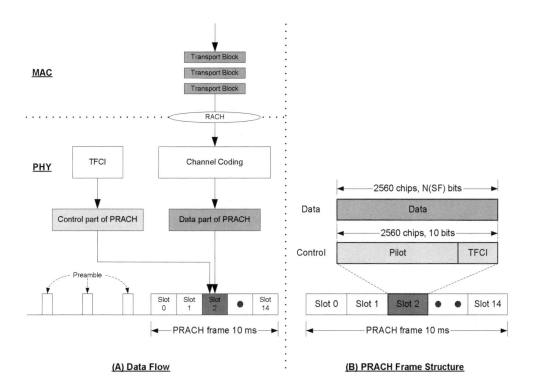

Figure 11.20 RACH Mapping Process and PRACH Frame Structure

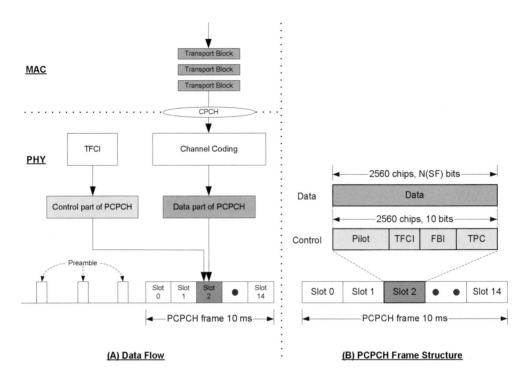

Figure 11.21 CPCH Mapping Process and PCPCH Frame Structure

The size of the preamble part of RACH and CPCH is 4096 chips and consists of 256 repetitions of a 16-chip signature. There are a maximum of 16 16-chip signatures available. The preamble part is coded with a scrambling code (Gold Code).

PRACH's control part deploys a channelization code with SF = 256 while PRACH's data part deploys a channelization code with SF ranging from 256 to 32.

PCPCH's control part deploys a fixed channelization code with SF = 256 while PCPCH's data part deploys a channelization code with SF ranging from 256 to 4.

11.3.3.1   Random Access Uplink Data Transmission Procedure (for RACH)

The Random Access Uplink Data Transmission is based on slotted-Aloha with fast acquisition indication. The data are transmitted at the beginning of an access slot. The access slots are spaced 5120 chips apart. There are 15 access slots spanning on two radio frames (2 x 10 ms = 20 ms) as shown in Figure 11.22. The Acquisition Indicator Channel (AICH) is a physical channel used for acquisition indication. AICH is defined at the base station whereas PRACH is defined at the mobile station. AICH and PRACH access slots are not aligned.

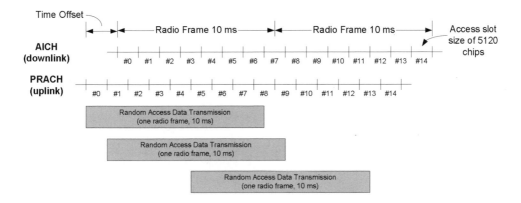

Figure 11.22 RACH Access Slots

As shown in Figure 11.23, RACH procedure includes the following steps:

- A mobile station retrieves the available RACH subchannel, scrambling codes, and signatures from BCH.
- The mobile station selects a RACH subchannel randomly and also further selects a signature from the pool of available signatures.
- RACH preamble that contains the selected signature is transmitted on a randomly selected access slot.
- If the mobile station does not receive the AICH preamble from the base station after the RACH preamble time out, the mobile station will retransmit the RACH preamble with higher transmission power on the next available access slot. If the number of retransmissions reaches the maximum number of retransmissions, the RACH procedure stops and notifies the upper layer.

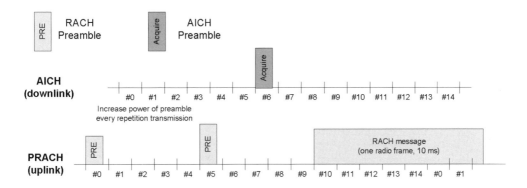

Figure 11.23 RACH Procedure

- If the mobile station receives the AICH preamble with NACK (negative acknowledgment), meaning that the selected signature is already used by another station, the RACH procedure stops and notifies the upper layer.
- If the mobile station receives the AICH preamble with ACK (positive acknowledgment), the mobile station transmits RACH messages with a duration of 10 or 20 ms.
- The backoff mechanism is controlled by the upper layer.

11.3.3.2   Random Access Uplink Data Transmission Procedure (for CPCH)
Random Access Uplink Data Transmission for CPCH is similar to that for RACH. The difference is that CPCH uses AP-AICH and CD/CA-ICH instead of AICH. In addition, CPCH includes a collision detection mechanism (CD) to reduce the probability of collision. The access slots are similar to those for RACH (see Figure 11.24). AP-AICH and CD/CA-ICH are defined at the base station whereas PCPCH is defined at the mobile station. AP-AICH and CD/CA-ICH access slots align but do not align with PCPCH access slots.

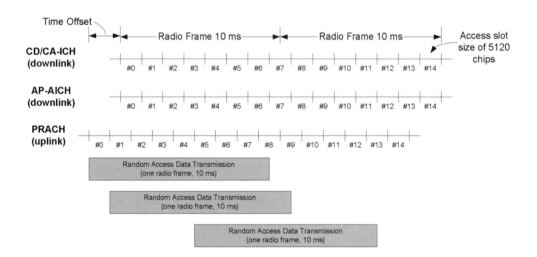

Figure 11.24 CPCH Access Slots

As shown in Figure 11.25 the CPCH procedure includes the following steps:

- The procedure is similar to RACH procedure up until the mobile station receives AP-AICH preamble with ACK.
- The mobile station transmits Collision Detection (CD) preamble to the base station.
- The base station echoes back with CD indication. This handshaking process reduces the probability of collision.
- The mobile station transmits a CPCH message.

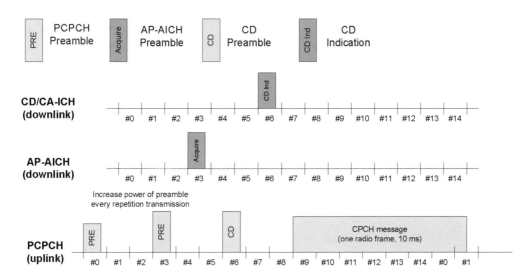

Figure 11.25 CPCH Procedure

11.4 Media Access Control (MAC)

The MAC provides the mapping process between logical and transport channels. Before we describe the MAC architecture, we introduce the logical channels.

11.4.1 Logical Channels

Logical channels (see Figure 11.26) are channels that relate to the content and what kind of data are transmitted through the radio interface. The Logical Channels are categorized into two types based on their content: Control Channels, which contain control information, and Traffic Channels, which contain user data.

The Control Channels (CCHs) include the following:

- Broadcast Control Channel (BCCH) is a downlink channel for broadcasting system control information.
- Paging Control Channel (PCCH) is a downlink channel for paging information. This channel is used when the network does not know the location cell of the mobile station, or when the mobile station is in the cell connected state using sleep mode procedures.
- Common Control Channel (CCCH) is a bidirectional channel for transmitting control information between the network and the mobile station. This channel is commonly used by a mobile station with no RRC connection with the network and by mobile stations using common transport channels when accessing a new cell after cell reselection.

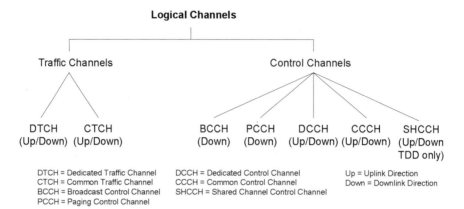

Figure 11.26 Logical Channels

- Dedicated Control Channel (DCCH) is a point-to-point bidirectional channel that transmits dedicated control information between the mobile station and the network. This channel is established through the RRC connection setup procedure.
- Shared Channel Control Channel (SHCCH) is a bidirectional channel that transmits control information for uplink and downlink shared channels between the network and the mobile stations. (*Note: This channel is used in TDD mode only.*)

The Traffic Channels (TCHs) include the following:

- Dedicated Traffic Channel (DTCH) is a point-to-point channel dedicated to a single mobile station for the transfer of user information. A DTCH can exist in both uplink and downlink.
- Common Traffic Channel (CTCH) is a point-to-multipoint unidirectional channel for transfer of dedicated user information for all or a group of specified mobile stations.

The logical channels can be mapped onto the transport channels in the MAC layer as shown in Figure 11.27.

11.4.2 MAC Architecture

Figure 11.28 presents the mobile station MAC architecture (Rel-5) and Figure 11.29 presents UTRAN MAC architecture. The MAC consists of the following modules: MAC-d, MAC-c/ch and MAC-b. In addition, UMTS Release 5 introduced a new MAC-hs module that supports HSDPA's high-speed mode.

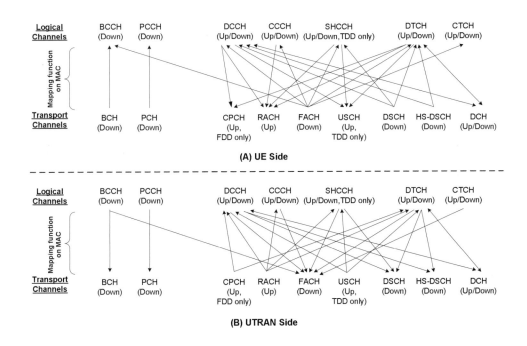

Figure 11.27 Logical Channels Mapped onto Transport Channels

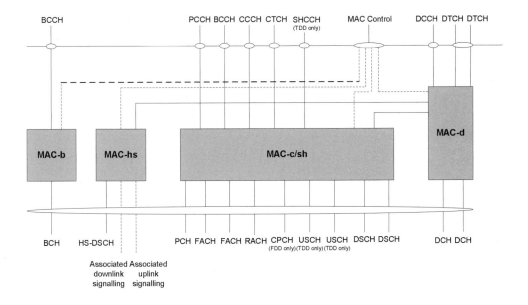

Figure 11.28 Mobile Station MAC Architecture

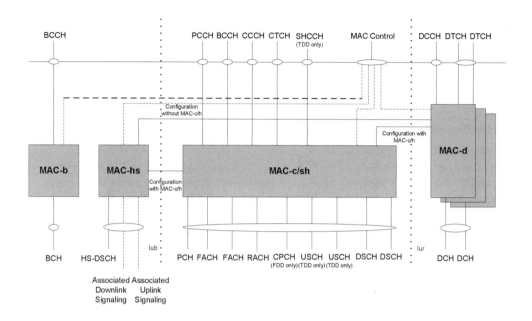

Figure 11.29 UTRAN Side MAC Architecture

The MAC executes the following functions:

- Maps between the logical and transport channels.
- Selects the appropriate Transport Format, defined in Section 11.3, from the Transport Format Combination Set assigned by the RRC for each Transport Channel depending on the instantaneous source rate. The Transport Format (which includes the Transport Block Size, the Transport Block Set Size, and the Transport Time Interval) accommodates the source bandwidth demand.
- Prioritizes between the data flows of a mobile station. Priority considerations can be achieved by selecting a Transport Format Combination for which the high-priority data are mapped into a "high bit rate" Transport Format, while lower priority data are mapped into a "low bit rate" Transport Format.
- Prioritizes between mobile stations via dynamic scheduling. The MAC handles priorities on the common and shared transport channels. For dedicated transport channels, the equivalent of the dynamic scheduling function is included as part of the RRC.
- Identifies mobile stations on common transport channels via in-band identification (i.e., UE ID).
- Multiplexes and demultiplexes upper layer packets into/from transport blocks delivered to/from the physical layer on common transport channels and dedicated transport channels.
- Measures traffic volume on the logical channels and reports it to the RRC. The RRC may switch transport channels based on these measurements.

- Switches between common and dedicated transport channels based on the switching decision made by the RRC.
- Encrypts data for security purposes.
- Accesses RACH and CPCH based on Service Classes. The RACH resources (access slots and preamble signatures for FDD, timeslot and channelization code for TDD) and CPCH resources (access slots and preamble signatures for FDD only) may be divided between different Access Service Classes (ASCs) in order to provide different priorities of RACH and CPCH. This MAC function indicates the appropriate backoff to the physical layer for the RACH and CPCH associated with the MAC packet transfer.
- Executes HARQ (Hybrid Automatic Repeat Request) functionality for HS-DSCH transmission to ensure reception of all packets. This function is carried by part of the MAC referred to as the MAC-hs.
- Delivers and assembles higher layer packets on HS-DSCH.

The execution of the aforementioned functions is undertaken by the following MAC entities: MAC-b for broadcast channels, MAC-c/sh for common channels, MAC-d for dedicated channels, and MAC-hs for the high-speed downlink shared channel. These MAC entities have different functionality at the mobile station and at the UTRAN.

11.4.2.1   MAC-b
MAC-b maps the BCCH logical channel to the BCH transport channel.

11.4.2.2   MAC-c/sh
MAC-c/sh handles the following transport channels: paging channel (PCH), forward access channel (FACH), random access channel (RACH), uplink common packet channel (CPCH) that exists only in FDD mode, downlink shared channel (DSCH), and uplink shared channel (USCH) that exists only in TDD mode. The mapping of logical channels on transport channels depends on the multiplexing that is configured by the RRC.

In the downlink, if the logical channels of the dedicated type are mapped to the common transport channels, MAC-d receives the data from MAC-c/sh or MAC-hs via the illustrated connection between the functional entities.

In the uplink, if the logical channels of the dedicated type are mapped to the common transport channels, MAC-d submits the data to MAC-c/sh via the illustrated connection between the functional entities.

The MAC controls the timing of RACH and CPCH transmissions. The physical RACH resources may be divided between eight different ASCs in order to provide different priorities of RACH usage. When the MAC determines that the packet has not been received based on the ACK, the MAC initiates retransmissions and backoff timing procedures before trying to retransmit. The number of retransmissions and the backoff timings are based on ASC considerations.

11.4.2.3   MAC-d
MAC-d handles the dedicated transport channel (DCH) to MAC-c/sh and MAC-hs.

11.4.2.4   MAC-hs

MAC-hs handles the high-speed downlink shared channel (HS-DSCH). It is a new entity introduced in Rel-5 that supports HSDPA (High-Speed Downlink Packet Access).

11.4.3 Examples of MAC Data Transmission

In this section we describe a number of examples of the MAC data transmission (Figure 11.30). The logical channel for user data is DTCH, which can be mapped to a dedicated transport channel (i.e., DCH), or common transport channel (i.e., RACH and CPCH for uplink, FACH for downlink), or shared transport channel (i.e., USCH for uplink, DSCH for downlink). If DTCH is mapped to a dedicated transport channel, the data travel through the MAC-d module, whereas if DTCH is mapped to a common transport channel or shared transport channel, the data travel through the MAC-d and MAC-c/sh modules. For HS-DSCH, the data travel through the MAC-d and MAC-hs modules. Figure 11.30A shows the uplink data flow when DTCH is mapped to RACH. The diagram presents only the channels involved in this flow. The data flow is presented in bold lines. Figure 11.30B illustrates the packet format at each layer.

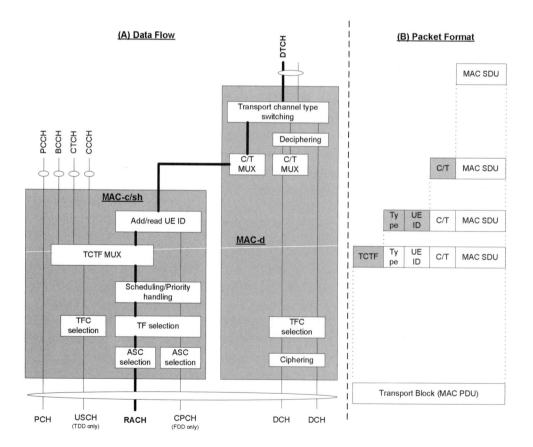

Figure 11.30 Example of MAC Data Transmission

The MAC SDU from DTCH visits the following modules sequentially:

1. Transport Channel Type Switching module, which decides to which transport channels to map it. The mapping is based on the traffic characteristics and QoS requirements (RACH for this example).

2. C/T MUX module, which is used when multiple dedicated logical channels are mapped to a transport channel. C/T header is added to the SDU to provide an identification of which logical channel the packet belongs to.

3. MAC-c/sh Add/Read UE ID module, which appends to the packet the UE ID header that includes the mobile station identification. The UE Type header (2 bits) provides information of what kind of UE ID is used (i.e., C-RNTI 16 bits or U-RNTI 32 bits).

4. TCTF MUX module, where the packet is multiplexed with packets from other common logical channels. TCTF header is also added to the packet, which includes the identification of the logical channel the packet belongs to.

5. Scheduling/Priority module, which provides priority services.

6. Transport Format (TF) Selection module, which selects a transport format from the transport format combination set.

7. ASC Selection module, which performs the priority random access procedure.

11.5 Data Link Layer Protocols (RLC, PDCP, and BMC)

11.5.1 Radio Link Control (RLC)

The RLC transfers data from the upper layers in transparent or non-transparent mode. In transparent mode upper layer packets are transferred without adding any protocol information. In non-transparent mode the RLC will segment and reassemble packets and may concatenate and pad packets to adjust the packets to the transport format. In addition, the RLC can transfer data in either unacknowledged or acknowledged mode. In unacknowledged data transfer, the RLC transmits upper layer packets without guaranteeing delivery to addressee. In acknowledged data transfer, the RLC guarantees delivery to the addressee. When the packet cannot be transmitted, the sender is being notified. The RLC uses sequence numbers to identify in-sequence and out-of-sequence delivery to allow resequencing of packets. In case errors are identified, the RLC requests retransmissions from the sender. The RLC maintains QoS as defined by the upper layers by providing different levels of service by means such as 1) notifying the upper layer of errors that cannot be resolved by the RLC and 2) adjusting the maximum number of retransmissions according to QoS delay requirements.

11.5.2 Packet Data Convergence Protocol (PDCP)

PDCP executes header compression and decompression of IP data streams (e.g., TCP/IP and Real Time Protocol (RTP)/UDP/IP headers) at the transmitting and receiving entity. It transfers user data from the upper layers and forwards the data to the RLC layer and vice versa and maintains PDCP sequence numbers in some cases.

11.5.3 Broadcast/Multicast Control (BMC)

BMC supports broadcast/multicast transmission services in the user plane for common user data in unacknowledged mode. The BMC stores broadcast messages for scheduled transmission and transmits these messages at the scheduled time. In the mobile station side the BMC also evaluates the scheduled messages to indicate the scheduling parameters to the RRC. The BMS also monitors traffic volume, calculates required resources and requests from the RRC these resources on CTCH and FACH. The BMS in the mobile station receives these broadcast messages and delivers error-free messages to the upper layers.

11.6 Radio Resource Control (RRC)

RRC executes the control plane signaling of the network layer between the mobile stations and the UTRAN network.

The RRC performs the following functions:

- Establishes, re-establishes, maintains, and releases RRC connections between the mobile station and UTRAN. The establishment of an RRC connection is initiated by a request from the mobile station higher layers. Multiple links can be established to a mobile station simultaneously. In this process, the RRC performs admission control and selects parameters describing the radio link based on information from higher layers. The RRC allocates the radio resource on the uplink and downlink so that the mobile station and UTRAN can communicate with the required QoS. The release of an RRC connection can be initiated by a request from higher layers or by the RRC layer itself in case of RRC connection failure. In case of connection loss, the RRC may re-establish the RRC connection. In case of RRC connection failure, the RRC releases resources associated with the RRC connection.
- Signals the allocation of radio resources to the mobile station.
- Evaluates, decides, and executes RRC connection mobility functions such as handover, preparation of handover to GSM or other systems, cell reselection, and cell/paging area update procedures.
- Ensures and controls that requested QoS parameters are met. In this capacity, the RRC allocates a sufficient number of radio resources.
- Controls measurements done by the mobile station—that is, it decides what to measure, when to measure, and how to report. It also reports these measurements to the network.
- Selects and reselects a cell based on idle mode measurements and cell selection criteria.
- Broadcasts information provided by the network to all mobile stations. For example, RRC may broadcast location service area information related to some specific cells. The RRC configures the BMC for cell broadcast services and allocates resources for this broadcasting.
- Performs other functions related to power control, encryption, and integrity protection.

The interaction between the mobile station's RRC and the UTRAN's RRC in the control plane can be viewed in Figure 11.31. The RRC also controls the radio resource through the RLC, MAC, and PHY layers.

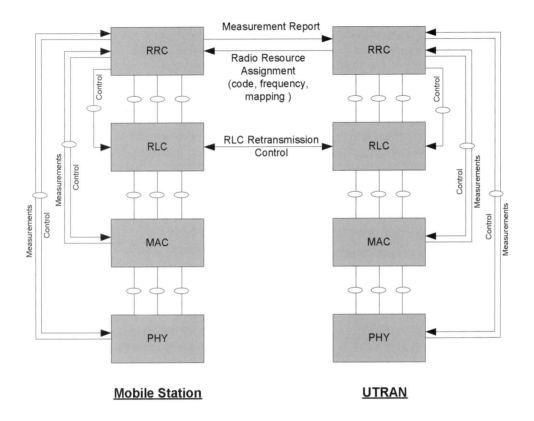

Mobile Station   **UTRAN**

Figure 11.31 RRC Interaction between UTRAN and the Mobile Station

11.6.1 RRC States

The mobile station can operate in either idle mode or connected mode. During startup, a mobile station performs the cell search procedure by scanning for the broadcast channel (BCH). After a cell is discovered and the mobile station decides to join the cell (sometimes called "camp on a cell"), the mobile station will be in idle mode and will keep listening to the BCH. In order to communicate with the UTRAN, the mobile station is required to establish an RRC connection with the UTRAN and changes to connected mode. One mobile station can have at most one RRC connection. As shown in Figure 11.32, in connected mode, the RRC is in one of the following four states: CELL_DCH, CELL_FACH, URA_PCH, and CELL_PCH. Each RRC state reflects the physical channels allocated to the mobile station.

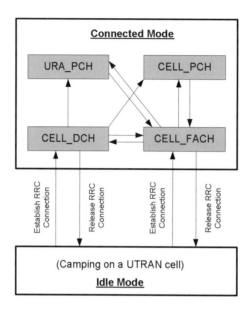

Figure 11.32 Mobile Station Modes and RRC States

In CELL_DCH RRC state, the dedicated channel is allocated. The RRC 1) reads system information broadcasted on FACH, 2) performs measurements, 3) selects and configures the radio links and multiplexing options applicable for the transport channels, and 4) acts upon RRC messages received on the control channels.

In CELL_FACH RRC state, the RACH and FACH are allocated. The mobile station 1) maintains up-to-date system information as broadcasted by the serving cell, 2) performs cell reselection processes, 3) performs measurements, 4) selects and configures radio links and multiplexing options applicable for the transport channels, and 5) acts upon RRC messages received on the control channels. When in this state and the station is out of service, the RRC performs a cell selection process.

In the URA_PCH or CELL_PCH state, there is no physical channel allocated. The mobile station maintains up-to-date system information as broadcasted by the serving cell, performs cell reselection process, and performs a periodic search for a higher priority network. It also monitors the paging channels and performs measurements. When the station is out of service, the station conducts a cell selection process.

Figure 11.33 illustrates RRC states for a mobile station that can operate in dual mode (UMTS and GPRS).

11.6.2 RRC Procedures

The standard defines several RRC functions and procedures. In this subsection, we introduce only the connection establishment and the radio bearer establishment procedures.

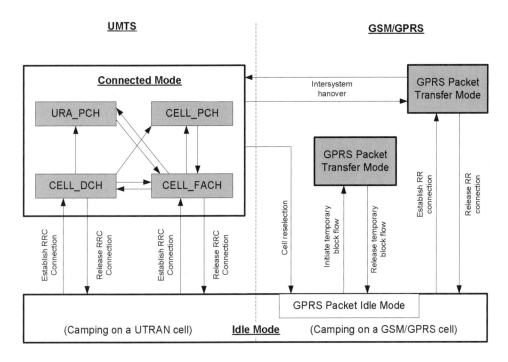

Figure 11.33 RRC States for a Dual Mode UMTS/GPRS Mobile Station

11.6.2.1   RRC Connection Establishment Procedure

The RRC connection establishment procedure (see Figure 11.34) is initiated when the upper layers in the mobile station need to establish a signaling connection and the mobile station is in idle
mode. First, the mobile station's RRC sends an "RRC Connection Request" message to
UTRAN's RRC through CCCH/RACH. Then, UTRAN's RRC replies with "RRC Connection
Setup" message through CCCH/FACH. If the RRC Connection Setup message includes a dedicated channel assignment, the mobile station will be in connected mode in CELL_DCH state.
On the other hand, if the RRC Connection Setup message includes a common channel assignment, the mobile station will be in connected mode in CELL_FACH state. Finally, the mobile
station confirms with a "RRC Connection Setup Complete" message through DCCH.

11.6.2.2   Radio Bearer Establishment Procedure

Figure 11.35 illustrates the radio bearer establishment procedure. The radio bearer establishment
procedure is initiated when the upper layers in the mobile station need to set up the transport and
physical channels to accommodate the traffic flow from the upper layers. This procedure starts
after the UTRAN performs the admission control and selects the parameters for the transport
(i.e., transport format set) and physical (i.e., channelization coding) channels.

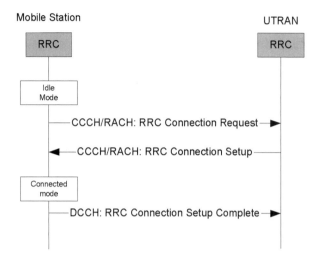

Figure 11.34 RRC Connection Establishment Procedure

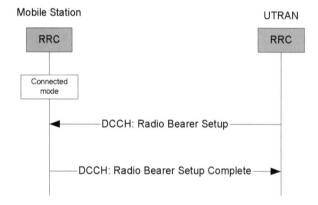

Figure 11.35 Radio Bearer Establishment Procedure

11.7 QoS Support

UMTS services adhere to the following QoS principles:

- Provide a finite set of QoS definitions and attributes that can be controlled. Their complexity should be reasonably low and the information involved should be kept low as well.
- Map between application requirements and UMTS services. The mapping should consider asymmetric uplink and downlink.
- Work with current QoS schemes and provide different QoS levels.

- Support session based QoS and allow multiple QoS streams per address.
- Manage QoS to yield efficient resource utilization.
- Modify QoS attributes when the session is ongoing and active.

The end-to-end QoS delivered to the user may be specified by Service Level Agreements (SLAs) between domains, parts of the network, and operators. The standard QoS end-to-end architecture is described in Figure 11.36. The bearer service defines the mechanisms (i.e., signaling, user plane transport, QoS management) that provide QoS support. Each service bearer relies on the service provided by the bearer service of the layer below. To enable QoS management based on SLAs across domains, consideration must be given to 1) the available bandwidth and QoS classes and how they map across domains, 2) policing and shaping requirements, 3) security mechanisms, and 4) reports of service usage, billing information, commercial information, and financial obligations in case of breech of contract, etc.

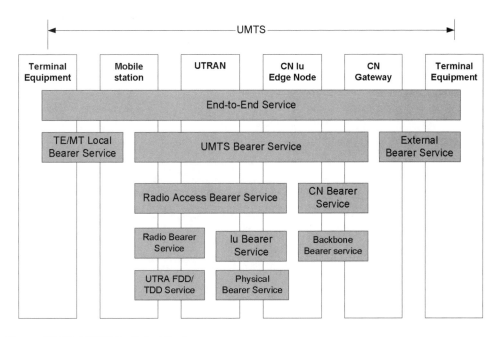

Figure 11.36 UMTS QoS Architecture

The standard provides an overview of the functionality needed to establish, modify, and maintain a UMTS link with a specific QoS. This functionality is divided into control and user planes. The standard provides examples of how such modules interact to request and commit QoS resources in conjunction with protocols such as Resource Reservation Protocol (RSVP). However, the standard does not provide the algorithms necessary to implement such functionality.

11.7.1 Control Plane QoS Management Functions

As shown in Figure 11.37, the control plane QoS management functions include several functions that are responsible for managing, translating, admitting, and controlling user requests and network resources.

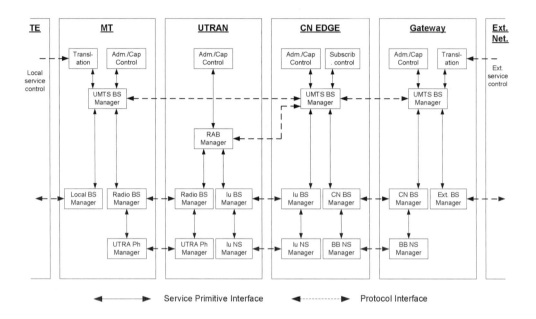

Figure 11.37 QoS Management Functions in the Control Plane. NS = network series, Ph = Physical, RAB (Radio Access Bearer).

The following functions in the control plane provide QoS management:

- *Service manager* coordinates the establishment, modification, and maintenance of services. It provides the user plane with QoS management functions. It also signals with peer service managers and uses services provided by other instances. The service manager may perform an attribute translation to request lower layer services and other control functions for service provisioning.
- *Translation function* maps between the internal service primitives for UMTS and other external protocols for service control. The translation includes the conversion between UMTS services and QoS attributes of the external network's service control protocol.
- *Admission/capability control* maintains information about all the network's available resources and all resources allocated to UMTS services. It determines for each UMTS service request or modification whether the required resources can be provided. When these resources are available they will be reserved for this service.
- *Subscription control* checks the administrative rights of the user to request the service with specified QoS attributes.

11.7.2 User Plane QoS Management Functions

The user plane QoS management functions are responsible for QoS signaling and monitoring of user data traffic. Such functions will ensure that the traffic is delivered within certain limits imposed by specific QoS attributes as negotiated with the UMTS network.

- *Mapping function* provides packets with the specific information required to receive the intended QoS attributes.
- *Classification function* assigns packets to the established services of the mobile station according to the related QoS attributes.
- *Resource manager* distributes the available resources between all services sharing the same resource according to the required QoS. For example, the resource manager can employ techniques such as scheduling, bandwidth management, and radio link power control.
- *Traffic conditioner* is a module that provides the traffic conformance with the negotiated QoS attributes. To achieve this goal, the traffic conditioner deploys traffic policing and/or traffic shaping mechanisms to all traffic. The traffic policing mechanism monitors the traffic's actual QoS attributes and compares them to the negotiated QoS attributes. If these attributes do not match, the traffic policing mechanism marks the packets as non-conformant or drops them. The traffic shaping mechanism shapes traffic according to the negotiated QoS attributes. For a more detailed description of traffic policing and traffic shaping mechanisms see Chapter 3.

An example of data flow within these various user plane QoS functions is shown in Figure 11.38.

11.7.3 QoS Classes

UMTS defines four different QoS classes: Conversational, Streaming, Interactive, and Background.

The main difference between these QoS classes is their sensitivity to time delay. The conversational class, for example telephony, is the most time delay sensitive while the background class is the least time delay sensitive.

The conversational and streaming classes carry real-time traffic flows. The conversational class is used for applications such as telephony (e.g., GSM), voice over IP and video conferencing applications. Real-time conversation is always performed between peers (or groups) of live (human) end users. Since the maximum transfer delay is dictated by the human perception of video and audio conversation, the limit for acceptable transfer delay is very strict. Failure to provide low enough transfer delay will result in unacceptable low quality.

The streaming class includes applications such as real-time video and audio streams. The highest acceptable delay variation over the transmission media is given by the capability of the time alignment function of the application.

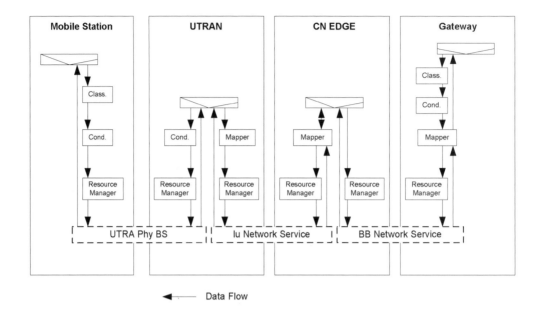

Data Flow

Figure 11.38 UMTS User Plane QoS Management Functions

The interactive class is mainly used by applications that request data from remote servers such as web browsing, data retrieval, and server access. Interactive traffic is characterized by the request-response time delay pattern of the end user.

The background class is used by applications performed in the background, such as downloads of emails and files. Since these applications are delay insensitive, the overall expectation of time delay is minimal or nonexistent.

As shown in Table 11.3, each traffic class is described by a set of QoS parameters.

Table 11.3 UMTS Bearer Service Attributes

UMTS Bearer Service Attribute	Conversational Class	Streaming Class	Interactive Class	Background Class
Maximum bit rate (kbps)— upper limit of the bit rate with which the UMTS delivers packets	2048	2048	< 2048— Overhead	< 2048— Overhead
Guaranteed bit rate (kbps)— guaranteed number of bits delivered by UMTS within a period of time	< 2048	< 2048		

Table 11.3 UMTS Bearer Service Attributes (Continued)

UMTS Bearer Service Attribute	Conversational Class	Streaming Class	Interactive Class	Background Class
Delivery order—indicates whether the UMTS link should provide in-sequence packet delivery or not	Yes/No	Yes/No	Yes/No	Yes/No
Maximum packet size (octets)	1500 or 1502	1500 or 1502	1500 or 1502	1500 or 1502
Delivery of erroneous packets —indicates whether packets detected as erroneous are delivered or discarded	Yes/No	Yes/No	Yes/No	Yes/No
Residual BER—indicates the undetected bit error ratio in the delivered packets	$5*10^{-2}$, 10^{-2}, $5*10^{-3}$, 10^{-3}, 10^{-4}, 10^{-5}, 10^{-6}	$5*10^{-2}$, 10^{-2}, $5*10^{-3}$, 10^{-3}, 10^{-4}, 10^{-5}, 10^{-6}	$4*10^{-3}$, 10^{-5}, $6*10^{-8}$	$4*10^{-3}$, 10^{-5}, $6*10^{-8}$
Packet error ratio—indicates the fraction of packets lost or detected as erroneous	10^{-2}, $7*10^{-3}$, 10^{-3}, 10^{-4}, 10^{-5}	10^{-1}, 10^{-2}, $7*10^{-3}$, 10^{-3}, 10^{-4}, 10^{-5}	10^{-3}, 10^{-4}, 10^{-6}	10^{-3}, 10^{-4}, 10^{-6}
Transfer delay (ms)—indicates the maximum delay for 95th percentile of delay distribution for all delivered packets during the lifetime of the communication link	100—Maximum value	250—Maximum value		
Traffic handling priority—differentiates and prioritizes between traffic with different QoS requirements within a traffic class			Under work by 3GPP	
Allocation/Retention priority—differentiates and prioritizes between traffic classes	Under work by 3GPP	Under work by 3GPP	Under work by 3GPP	Under work by 3GPP

Table 11.3 UMTS Bearer Service Attributes (Continued)

UMTS Bearer Service Attribute	Conversational Class	Streaming Class	Interactive Class	Background Class
Source statistic descriptor—specifies source characteristics	Speech/ unknown	Speech/ unknown		
SDU format information—indicates the fraction of packets lost or detected as erroneous	Under work by 3GPP	Under work by 3GPP		

11.7.4 QoS Mechanisms in the Radio Bearer Service

Figure 11.39 depicts a simplified diagram of the radio bearer key QoS mechanisms that reside in the RRC, MAC, and PHY layers. The RRC, which plays the center role for QoS management, manages the signaling between the mobile station and the UTRAN. Furthermore, the RRC can control the MAC and PHY layers.

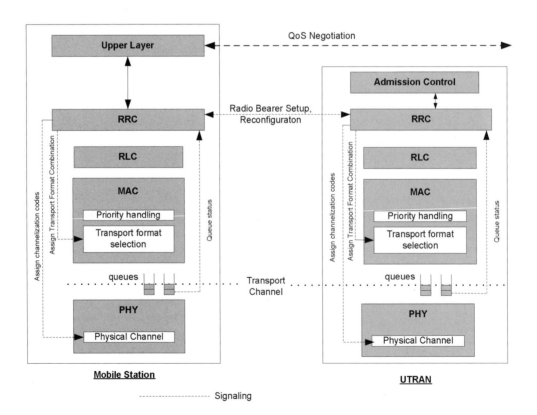

Figure 11.39 QoS Mechanisms in Radio Bearer

First, an application performs QoS negotiation with the UMTS core network in the upper layer. The upper layer service attributes are mapped into the radio access bearer (RAB) service attributes. The RRCs of the mobile station and UTRAN perform the radio bearer setup and reconfiguration procedures. During these procedures the UTRAN's RRC consults with UTRAN's admission control module in order to check if there are enough resources and make sure that the new connection does not interfere and does not degrade the existing services. In case the application is admitted, the RRC allocates bandwidth through channelization code assignment in the physical layer and transport format combination in the MAC layer. Furthermore, the MAC also contains a priority handling mechanism.

Next, we describe the following aspects of the radio bearer QoS mechanisms: bandwidth allocation and priority handling.

11.7.4.1   Bandwidth Allocation

After a new application performs the QoS negotiation, the RRC establishes the radio bearer services which define the transport and physical channels. The transport channel is controlled by the MAC while the physical channel is controlled by the PHY.

As discussed in Section 11.3.1, the parameters that describe the transport channel are included in the Transport Format: transport block size, transport block set size, and transport time interval. Therefore, these parameters will define the bandwidth allocation.

The physical channel parameter is the channelization code. As described in Section 11.3, the bandwidth allocation can be changed by changing the spread factor (SF). The smaller the spread factor code, the larger the bandwidth (or data rate).

The queue size of the transport layer (at the uplink and at the downlink) is used to determine the application's dynamic bandwidth demand. When this queue grows beyond a certain threshold, the RRC will perform radio bearer reconfiguration by adjusting the physical channel configuration (i.e., adjust channelization code) and the transport channel configuration (i.e., adjust transport format and channelization code). Figures 11.40 to 11.42 show such reconfiguration examples for different scenarios.

11.7.4.2   Priority Handling

Priority handing is located in the MAC-c/sh as shown in Figure 11.30. The traffic with the highest priority receives service first. The priority value is related to the Transport Format selection. A high priority value is associated with a Transport Format that provides high data rate while a low priority value is associated with a Transport Format that provides low data rate.

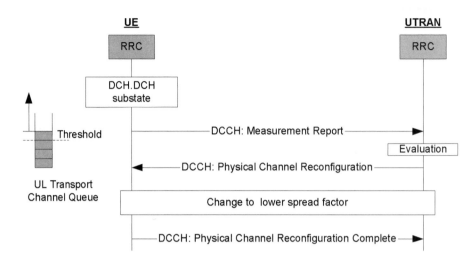

Figure 11.40 Increase of Uplink Bandwidth Using Physical Channel Reconfiguration

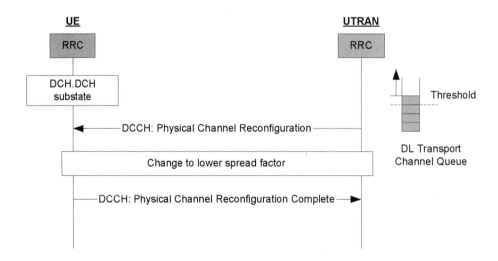

Figure 11.41 Increase of Downlink Bandwidth Using Physical Channel Reconfiguration

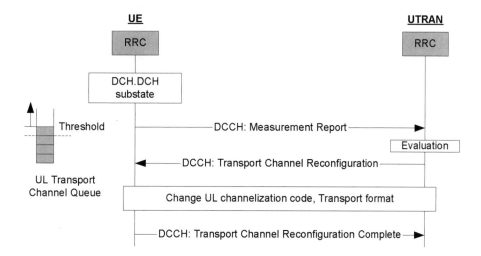

Figure 11.42 Increase of Uplink Bandwidth Using Transport Channel Reconfiguration

CHAPTER 12

cdma2000

12.1 Introduction

cdma2000 is a 3G technology whose standards are specified by the Third Generation Partnership Project 2 (3GPP2). 3GPP2 is a collaborative organization that includes members from all over the world, with a majority of members from North America and Asia. 3GPP2 is part of IMT-2000 (International Mobile Telecommunication–2000) effort that is designed to provide high-speed communication with high-quality multimedia services and global roaming support.

Figure 12.1 shows the evolution of the cdma2000 standard. cdmaOne, or IS-95, was influenced by various standards such as Advanced Mobile Phone Service (AMPS), IS-136 TDMA, and GSM. IS-95 has two revisions: IS-95A and IS-95B. IS-95A is a 2G technology with a data rate of 14.4 kbps, while IS-95B is 2.5G technology with a data rate of 115 kbps. The 3G cdma2000 continued to evolve and support IS-95.

cdma2000, which is also referred to as IMT2000-Mc (IMT-CDMA Multicarrier), has been divided into the following two phases: Phase 1 cdma2000 1x (sometimes called cdma2000 1xRTT) and Phase 2 cdma2000 3x (sometimes called cdma2000 3xRTT). cdma2000 1x deploys a single radio frequency carrier (1.25 MHz bandwidth) and delivers 307 kbps in a mobile environment, whereas cdma2000 3x deploys multicarrier technology (i.e., a multiple of 1.25 MHz bandwidth carriers) and delivers speeds of up to 2 Mbps. cdma2000 1x continued to evolve to cdma2000 1x Evolution (1xEV). cdma2000 1xEV is backwards compatible with cdma2000 1x and cdmaOne.

cdma2000 1xEV utilizes Frequency Division Duplex (FDD), where the uplink (from the mobile station to the base station, also called reverse link) and downlink (from the base station to the mobile station, also called forward link) operate on two separate frequency bands.

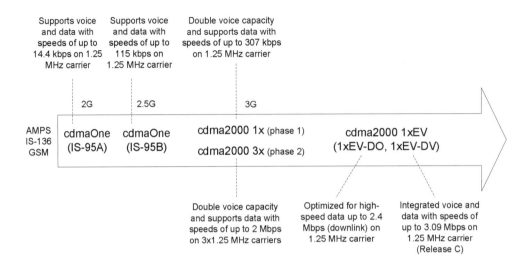

Figure 12.1 cdma2000 Evolution

cdma2000 1xEV has two variations: cdma2000 1xEV-DO (Data Only) optimized for high-speed data transmission (up to 2.4 Mbps) and cdma2000 1xEV-DV (Data and Voice), which supports both data and voice with speeds of up to 3.09 Mbps.

cdma2000 has been introduced in several releases. In this book we describe cdma2000 Release C. We focus on the following 3GPP2 documents:

- 3GPP2 C.S0001-C Introduction to cdma2000 Standards for Spread Spectrum Systems Release C
- 3GPP2 C.S0002-C Physical Layer Standard for cdma2000 Spread Spectrum Systems Release C
- 3GPP2 C.S0003-C Medium Access Control (MAC) Standard for cdma2000 Spread Spectrum Systems Release C
- 3GPP2 C.S0004-C Signaling Link Access Control (LAC) Standard for cdma2000 Spread Spectrum Systems Release C
- 3GPP2 C.S0005-C Upper Layer (Layer 3) Signaling Standard for cdma2000 Spread Spectrum Systems Release C

12.2 cdma2000 Architecture

Figure 12.2 illustrates the basic network architecture of cdmaOne (IS-95) and cdma2000. A mobile station connects to the external network (i.e., Public Switched Telephone Network [PSTN] or Internet) through the base station and core network. The base station consists of two entities: Base Transceiver Station (BTS) and Base Station Controller (BSC). BTS provides the communication services within its coverage area or cell. BSC manages the call handoff and radio resources of each BTS. In the IS-95 system, the BSC connects to the Mobile Switching Center (MSC) accommodating both voice and data traffic. The voice traffic is routed through the MSC to the external telephone networks (i.e., PSTN), whereas the data traffic is routed to the external data network (i.e., Internet) through the Interworking Function (IWF). IWF provides the access point to the Internet. In a cdma2000 system, there are separate links between the base station and the core network. The link between the BSC and the MSC accommodates voice services while the link between the BSC and the Packet Data Service Node (PDSN) accommodates data services. PDSN supports, establishes, maintains, and terminates IP sessions for the mobile station. The AAA module is responsible for authorization and accounting.

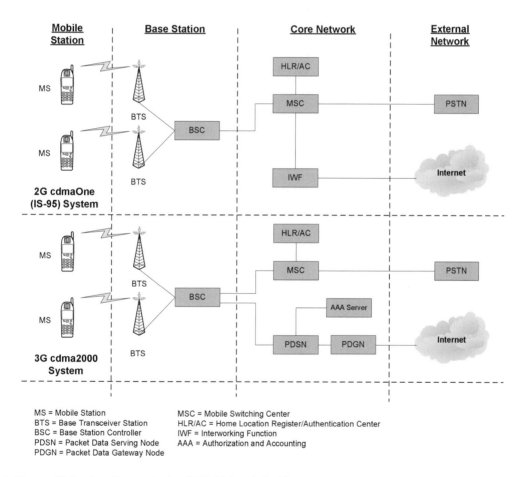

MS = Mobile Station   MSC = Mobile Switching Center
BTS = Base Transceiver Station   HLR/AC = Home Location Register/Authentication Center
BSC = Base Station Controller   IWF = Interworking Function
PDSN = Packet Data Serving Node   AAA = Authorization and Accounting
PDGN = Packet Data Gateway Node

Figure 12.2 cdmaOne and cdma2000 Network Architecture

12.2.1 cdma2000 Air Interface Protocol Architecture

Figure 12.3 illustrates cdma2000 air interface protocol architecture. The air interface protocol architecture is divided into layers corresponding to the Open Systems Interconnection (OSI) protocol layers: 1) Layer 1 or physical layer; 2) Layer 2, which includes the Link Access Control (LAC) sublayer, the Media Access Control (MAC) sublayer, and the Forward Packet Data Channel (F-PDCH) Control Function sublayer; and 3) Layers 3 to 7, which include the upper layer signaling and the voice and data services originated from the users' applications. The F-PDCH Control Function module newly introduced in Revision C contains several key features such as Adaptive Modulation and Coding scheme (AMC) and Hybrid Automatic Repeat request (HARQ). There are also signaling interfaces between the upper layer and the physical layer.

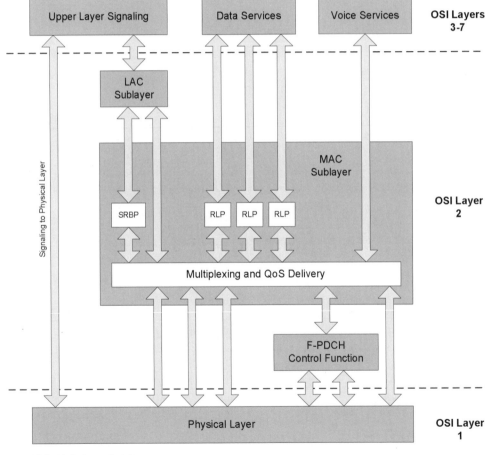

MAC = Media Access Control
LAC = Link Access Control
PDCH = Packet Data Channel Control Function
RLP = Radio Link Protocol
SRBP = Signalling Radio Burst Protocol

Figure 12.3 cdma2000 Air Interface Protocol Architecture

A more detailed description of cdma2000 air interface protocol architecture at the mobile station side is shown in Figure 12.4.

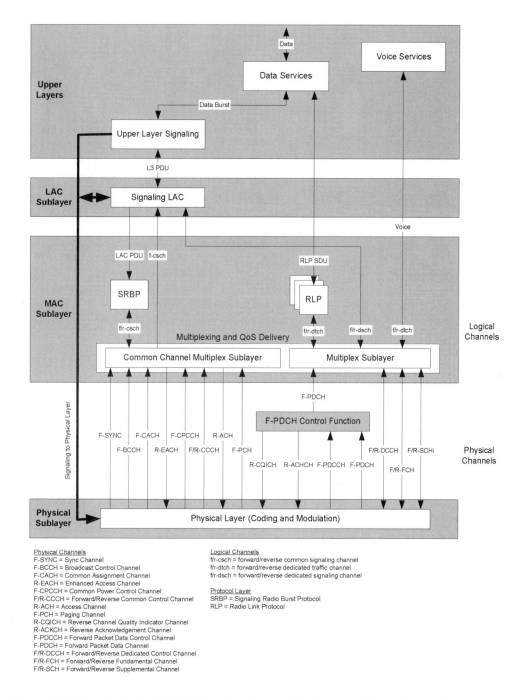

Figure 12.4 Detailed cdma2000 Air Interface Protocol Architecture (Mobile Station Side)

cdma2000 is based on physical and logical channels that carry both data and control packets. The logical channel defines what type of information (i.e., data message, control message) is delivered. Packets from each logical channel are multiplexed and delivered in timely fashion by the MAC to the appropriate physical channels. The MAC delivers these packets based on the QoS requirements of each logical channel. The physical channel defines the radio transport channel based on the radio configuration, which is a combination of encoding, interleaving, frame size, and bit rate. The radio configuration of the physical channel is defined in the physical layer. As shown in Figure 12.4 the connection between the MAC and the physical layers indicates the physical channel to which the connection belongs.

There are two directions on both the logical and physical channels: forward direction (from the base station to the mobile station) and reverse direction (from the mobile station to the base station). These directions are included as the first letter (f = forward, r = reverse) of the logical and physical channel name. A logical channel name consists of three lower case letters and the suffix "ch" (channel). The first letter (i.e., f = forward and r = reverse) indicates the direction of the logical channel. "f" and "r" can be used together if the logical channel has both directions. The second letter (i.e., d = dedicated or c = common) indicates the type of logical channel (dedicate or common [share]). The third letter (i.e., t = traffic or s = signaling) indicates the contents. The first and second letters are separated by a hyphen. Examples of logical channels include f-dtch (forward dedicated traffic channel) and f/r-dsch (forward/reverse dedicated signaling channel). The physical channel name is written in capital letters. The first letter (i.e., F = forward or R = reverse) indicates the direction of the physical channel such as F-PDCH (Forward Packet Data Channel).

12.3 Physical Layer

cdma2000 uses an FDD Code Division Multiple Access (CDMA) network—that is, the transmission from the base station to the mobile station, referred to as forward traffic (i.e., downlink), is done on a different frequency than that of the traffic from the mobile station to the base station, referred to as reverse traffic (i.e., uplink) as shown in Figure 12.5. For each forward and reverse link, cdma2000 employs a combination of Time Division Multiplex (TDM) and Code Division Multiplex (CDM). For cdma2000 1x, the frequency channel width of both the forward and reverse links is 1.25 MHz. These frequency channels are separated by 45 MHz.

As shown in Table 12.1, cdma2000 is assigned various forward and reverse frequencies based on the country of application.

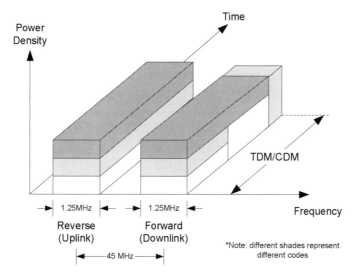

Figure 12.5 Channel Access

Table 12.1 cdma2000 Frequency Bands

Band Class	System	Transmit Frequency Band (MHz)	
		Mobile Station (Reverse Link)	**Base Station (Forward Link)**
0	North America Cellular	824-849	869-894
1	North America PCS	1850-1910	1930-1990
2	Total Access Communication System	872-915	917-960
3	Japan Total Access Communication System	887-925	832-870
4	Korean PCS	1750-1780	1840-1870
5	Nordic Mobile Telephone	411-484	421-494
6	IMT-2000	1920-1980	2110-2170

Table 12.1 cdma2000 Frequency Bands (Continued)

Band Class	System	Transmit Frequency Band (MHz)	
		Mobile Station (Reverse Link)	Base Station (Forward Link)
7	North America 700	776-794	746-764
8	1800 MHz	1710-1785	1805-1880
9	900 MHz	880-915	925-960

The physical layer delivers the packets received from the MAC layer to the physical channels. The physical layer defines a combination of frames, modulations, and codes used for each physical channel.

For channel access, the physical channel deploys CDMA, in which the information bits are spread by using a number of Pseudo-Noise (PN) chip signals (a multiple of 1.2288 Mcps). The cdma2000 Release C defines two spreading rates: Spreading Rate 1 and Spreading Rate 3. Spreading Rate 1 (referred to as 1x) uses 1.2288 Mcps PN chip signal spread over a 1.25 MHz channel. Spreading Rate 3 (referred to as 3x) has two spreading approaches. The first approach uses 1.2288 Mcps PN chip signal spreading over each of three 1.25 MHz channels (sometimes called Multicarrier Channel). This approach is used in forward CDMA channels. The other approach uses 3.6864 Mcps PN chip signal spreading over a 3.75 MHz channel. The latter approach is used in reverse CDMA channels. Figure 12.6 illustrates the Spreading Rate.

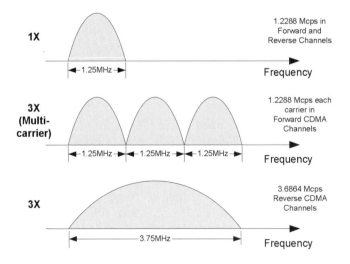

Figure 12.6 Spreading Rate

The CDMA involves two types of codes: long code and Walsh code. The long code (similar to the scrambling code in Universal Mobile Telecommunications System [UMTS]) uses PN sequence to scramble the forward and reverse CDMA channels. On the reverse channel, different long codes are used to identify transmissions from different mobile stations. On the forward channel, the phase of the long code is used to identify the base station transmissions. Unlike coding in UMTS, where the different scrambling codes are used to differentiate between transmissions from different base stations, in cdma2000 each base station uses the same long code but different phases. There are a total of 512 long code phases. The Walsh code (similar to the channelization code in UMTS) is used to identify the physical channels within a mobile station or within a base station. cdma2000 has a total of 128 Walsh codes. The basic concept of Walsh code, which is referred to as channelization code, is described in Chapter 11, Section 11.3.

cdma2000 also defines the radio configurations (RCs) for the forward and reverse CDMA channels. The radio configuration consists of a combination of physical layer parameters such as spreading modulation (i.e., BPSK, QPSK), spreading rate (i.e., 1x, 3x), forward error correction (i.e., 1/2 to 1/4 convolution code, 1/2 to 1/5 turbo code), and data rate. There are six radio configurations (RC1 to RC6) for the reverse CDMA channel and ten radio configurations (RC1 to RC10) for the forward CDMA channel.

12.3.1 Physical Channels

The standard defines the structure of the physical channel in frames. The frame structure varies on each type of physical channel with possible frame sizes of 1.25 ms, 2.5 ms, 5 ms, 10 ms, 20 ms, 40 ms, and 80 ms. The physical channels include the following:

- Forward/Reverse Fundamental Channel (F/R-FCH)
- Forward/Reverse Dedicated Control Channel (F/R-DCCH)
- Forward/Reverse Supplemental Code Channel (F/R-SCCH)
- Forward/Reverse Supplemental Channel (F/R-SCH)
- Paging Channel (F-PCH)
- Quick Paging Channel (F-QPCH)
- Access Channel (R-ACH)
- Forward/Reverse Common Control Channel (F/R-CCCH)
- Forward/Reverse Pilot Channel (F/R-PICH)
- Transmit Diversity Pilot Channel (F-TDPICH)
- Auxiliary Pilot Channel (F-APICH)
- Auxiliary Transmit Diversity Pilot Channel (F-ATDPICH)
- Sync Channel (F-SYNCH)
- Common Power Control Channel (F-CPCCH)
- Common Assignment Channel (F-CACH)
- Enhanced Access Channel (R-EACH)
- Broadcast Control Channel (F-BCCH)
- Forward Packet Data Channel (F-PDCH)
- Forward Packet Data Control Channel (F-PDCCH)

• Reverse Acknowledgment Channel (R-ACKCH)

• Reverse Channel Quality Indicator Channel (R-CQICH)

In the following subsection we introduce some of these physical channels.

12.3.1.1   Pilot Channels

There are pilot channels in both forward and reverse links. In the forward link, pilot channels include Forward Pilot Channel (F-PICH), Transmit Diversity Pilot Channel (F-TDPICH), Auxiliary Pilot Channel (F-APICH) and Auxiliary Transmit Diversity Pilot Channel (F-ATDPICH). Forward Pilot Channel is transmitted by the base station at all times. F-PICH helps the mobile station in the initial cell search process. When a new mobile station joins a cell, it detects the F-PICH transmitted by the base station. The mobile station also measures the forward signal strength from the pilot channel. F-TDPICH has the same function as F-PICH but it is used in transmit diversity. F-APICH and F-ATDPICH are used in beam forming antenna systems. The Reverse Pilot Channel (R-PICH) is only a pilot channel in the reverse link which helps the base station detect the mobile station. R-PICH also includes the Reverse Power Control Subchannnel which controls the power of the forward link.

12.3.1.2   Physical Channels for Voice and Data Traffic

Figure 12.7, which is a part of Figure 12.4, illustrates the physical channels for voice and data traffic. cdma2000 1x defines Forward/Reverse Fundamental Channel (F/R-FCH), Forward/Reverse Supplemental Channel (F/R-SCH), and Forward/Reverse Dedicated Control Channel (F/R-DCCH) to support voice and data traffic. In addition, cdma2000 1xEV introduces the F-PDCH Control Function, which contains new physical channels that support the high-speed forward data transmission. These new physical channels include Forward Packet Data Channel (F-PDCH), Forward Packet Data Control Channel (F-PDCCH), Reverse Acknowledgment Channel (R-ACKCH), and Reverse Channel Quality Indicator Channel (R-CQICH).

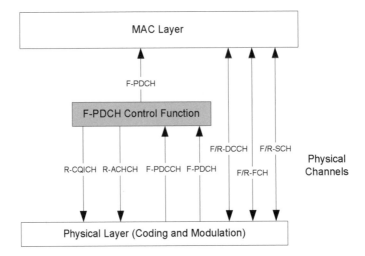

Figure 12.7 Physical Channels for Voice and Data Traffic (mobile station side)

The Forward/Reverse Fundamental Channel (F/R-FCH) can support voice, data, and signaling. The standard defines a flexible data rate that ranges from 750 bps to 14.4 kbps by varying the spreading rate and the associated set of frames for each radio configuration. The Forward/Reverse Dedicated Control Channel (F/R-DCCH) can be used for signaling or burst data transmissions. The Forward/Reverse Supplemental Channel (F/R-SCH) is defined to support high rate data services. F/R-SCH is scheduled dynamically on a frame-by-frame basis. F/R-SCH can deliver data rates up to 32 times the rates of an F/R-FCH. At the base station, the number of F-SCHs is limited by the available transmission power and Walsh codes. At the mobile station, the number of R-SCHs is limited to two.

Forward Packet Data Channel (F-PDCH) provides high-speed forward data transmission. A base station can support up to two F-PDCHs, and each F-PDCH transmits data packets to one mobile station at a time. A mobile station can support one F-PDCH at a time. Forward Packet Data Control Channel (F-PDCCH) provides the transmitting control information (i.e., users' MAC ID, packet size, subpacket ID) for the associated F-PDCH. F-PDCCH can also be used to broadcast the Walsh mask to the mobile stations. A base station can support up to two F-PDCCHs (corresponding to two F-PDCHs). When a mobile station detects its own MAC ID on a F-PDCCH, it will retrieve data packets from the associated F-PDCH. The high forward data rate can be achieved through Adaptive Modulation and Coding (AMC) scheme and flexible Time Division Multiplex/Code Division Multiplex (TDM/CDM). The AMC scheme includes a combination of a number of information bits (i.e., 386, 770, 1538, 2306, 3074 or 3842), the total frame duration (i.e., 1.25 ms., 2.5 ms, or 5 ms), modulation schemes (i.e., QPSK, 8-PSK or 16-QAM), and spreading codes. Table 12.2 shows F-DPCH possible data rates.

Table 12.2 F-DPCH Data Rates (kbps)

		F-PDCH Packet Size (bits) (information bits + 16 quality indicator bits + 6 turbo encoder tail bits)					
		408 bits	**792 bits**	**1560 bits**	**2328 bits**	**3096 bits**	**3864 bits**
Total Frame Duration (ms)	5 ms	82 kbps	158 kbps	312 kbps	466 kbps	619 kbps	773 kbps
	2.5 ms	163 kbps	317 kbps	624 kbps	931 kbps	1238 kbps	1546 kbps
	1.25 ms	326 kbps	634 kbps	1248 kbps	1862 kbps	2477 kbps	3091 kbps

The AMC scheme has flexible choices to achieve the different data rates. TDM/CDM provides packet scheduling of multiple F-PDCHs (up to two F-PDCHs transmitting at the same time) by varying the Walsh Code and scheduling. An example of a TDM/CDM assignment is shown in Figure 12.8 for four mobile stations (MS1 to MS4).

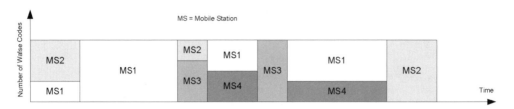

Figure 12.8 Example of a TDM/CDM Assignment

The Reverse Acknowledgment Channel (R-ACKCH) enables Automatic Repeat Request (ARQ) mechanism. After successfully receiving packets from the base station, the mobile station sends an acknowledgment to the base station through the R-ACKCH. The fact that the ARQ mechanism resides at the physical layer speeds up the retransmission process. Furthermore, instead of using the typical ARQ mechanism, cdma2000 1xEV employs a Hybrid ARQ that combines ARQ with Forward Error Correction (FEC). To improve the error correction performance and reduce the number of retransmission attempts, Hybrid ARQ performs FEC decoding on both erroneous packets from the previous transmission and packets from the current transmission.

Reverse Channel Quality Indicator Channel (R-CQICH) is used by the mobile station to indicate to the base station the channel quality measurements. The base station then uses this information to determine the appropriate spreading, modulation, frame size, and scheduling of F-PDCH. It allows the base station to efficiently manage the radio resources.

In summary, the voice and data packets are transmitted from the base station through F-FCH, F-DCCH, F-SCH, and F-PDCH (in case high-speed forward transmission is required). The voice and data packets are transmitted from the mobile station to the base station through F-FCH, F-DCCH, and F-SCH.

12.4 Media Access Control (MAC)

Figure 12.9 illustrates the MAC layer and its interactions with the LAC. cdma2000 supports a generalized model of multimedia services that allows simultaneous support for both voice and data traffic. cdma2000 also includes Quality of Service (QoS) control mechanisms that reside at the MAC layer. These mechanisms balance different QoS requirements of multiple simultaneous multimedia applications. The key functions of the MAC layer are 1) to receive the packets (i.e., voice, data) from the upper layers and 2) to schedule these packets to physical channels in a timely manner based on the connection QoS requirements. This process is done by the MAC through a Multiplexing and QoS Delivery module (see Figure 12.9). Release C standard also defines the signaling between the LAC and the MAC using service primitives. This signaling provides QoS information that can be used by the MAC. The standard also defines the following two modules: Signaling Radio Burst Protocol (SRBP) and Radio Link Protocol (RLP). SRBP is a connectionless protocol for signaling messages. RLP, which is a connection-oriented protocol, provides reasonably reliable transmission over the radio link using a negative-acknowledgment-based data delivery protocol.

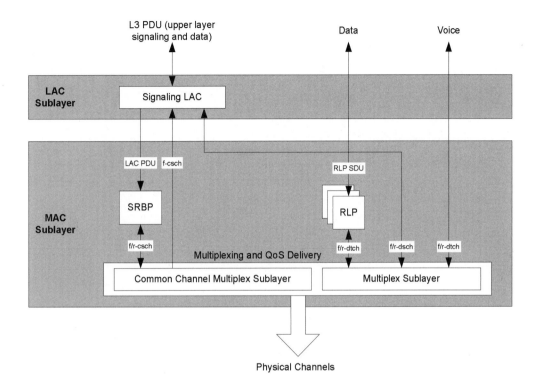

Figure 12.9 MAC Layer

We introduce four key service primitives used in the signaling procedure between the Signaling LAC and the MAC:

- *MAC-SDUReady.Request:* This primitive is sent from the signaling LAC to the MAC when there are packets waiting in the signaling LAC. The primitive includes the following parameters:

  - channel_type: the type of channel required—"5ms FCH/DCCH frame," "20ms FCH/DCCH frame," or "F-PDCH frame"
  - size: the size of packets (bits)
  - scheduling hint: indicates the relative priority service requirement

- *MAC-Availiability.Indication:* This primitive is sent from the MAC to the signaling LAC when the MAC multiplex module is ready to receive the packets from the upper layer and to transmit them to the physical layer. The primitive includes the following parameters:

  - channel_type: the type of channel allowed—"5ms FCH/DCCH frame," "20ms FCH/DCCH frame," or "F-PDCH frame"

- max_size: the maximum number of bits from the signaling LAC that can be fitted in the physical channel based on QoS requirements
- system_time: indicates the time when the physical layer will transmit the packets

- *MAC-Data.Request:* This primitive is sent from the signaling LAC to the MAC. It includes the type of channel required and a packet. The MAC multiplex module will fit this packet to the closest size data block determined by the rate set of the physical channel.
- *MAC-Data.Indication:* This primitive is sent from the MAC to the signaling LAC to indicate the packet transmission in the physical layer.

Figure 12.10 illustrates the signaling and packet transmission process (steps 1 to 5) at the MAC layer.

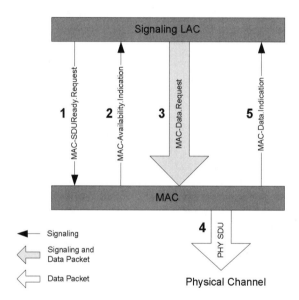

Figure 12.10 Signaling and Packet Transmission Process

12.4.1 Logical Channels

The following logical channels define what type of information (i.e., user data message, control message) is delivered:

- *Forward/Reverse Dedicated Traffic Channel (f/r-dtch):* A point-to-point logical channel that carries voice and data packets and transmits through a dedicated physical channel.

- *Forward/Reverse Dedicated Signaling Channel (f/r-dsch):* A point-to-point logical channel that carries signaling packets from the upper layer to a dedicated physical channel.
- *Forward/Reverse Common Signaling Channel (f/r-csch):* A point-to-multipoint logical channel that carries signaling packets from the upper layer to a common physical channel.

12.4.2 Multiplex and QoS Sublayers

The multiplex sublayer is responsible for transmitting and receiving packets from the physical layer. The multiplex sublayer receives information from the logical channels that comes from various sources (i.e., upper layer signaling, data, and voice services). The transmitting function, under QoS control, solicits information from various packets exchanged with the MAC. This information is used by the multiplex sublayer to determine the relative priority between the traffic supplied by signaling and other services. The exact manner for using this information to deliver over-the-air QoS is not specified by the cdma2000 standard.

12.5 Link Access Control (LAC)

The LAC, which includes several sublayers as depicted in Figure 12.11, performs the following functions:

- Delivery of packets with optional ARQ techniques to provide reliability
- Segmentation of packets to sizes suitable for the MAC and reassembly of such packets coming from the MAC
- Access control through authentication
- Address control to ensure delivery of packets based upon addresses that identify particular mobile stations
- Internal signaling, by exchanging notifications and data with the MAC and the supervisory and configuration entities, resulting from the processing of LAC Sublayer level information

12.6 QoS Support

cdma2000 Release C standard supports voice and high-speed data transmission. Each application has different QoS requirements. As we described in previous sections, cdma2000 defines several QoS mechanisms that reside at different protocol layers (i.e., physical layer, MAC layer, and upper layers). In this section we summarize the QoS mechanisms defined by the standard.

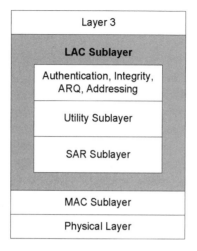

Figure 12.11 LAC Sublayer

12.6.1 Bandwidth Allocation

The standard defines Adaptive Modulation and Coding (AMC) and flexible TDM/CDM. Both techniques provide flexible tools that enable dynamic bandwidth allocation to match with the current traffic and channel conditions. The algorithm that determines the bandwidth allocation is not defined by the standard. The product developers need to develop these algorithms in order to optimize their network resources.

12.6.2 Packet Scheduling

Packet scheduling algorithms indicate when packets from a user or application are allowed to transmit. The standard defines the multiplex and QoS delivery sublayers in the MAC to support packet scheduling algorithms. Multiplex and QoS delivery sublayers can provide service differentiation among logical channels (i.e., voice, data, signaling). The standard also defines the service primitives between the signaling LAC and MAC that enable the transfer of QoS requirement information from the upper layers. The packet scheduling algorithms are not defined by the standard.

CHAPTER 13

Satellite Communication

13.1 Introduction

Communication satellites allow the establishment of communication worldwide in both populated and remote areas, where other ways of communication may be impossible (e.g., oceans and mountains).

Traditionally, satellites have been used for broadcasting television programs. In recent years, the satellite industry has evolved and has started to provide two-way communication. This evolution is possible due to technology, decreasing cost of user's equipment, and its availability in all world regions. These satellite systems are either proprietary or standardized. Hughes Network Systems with its Spaceway program and Gilat Satellite Networks are examples of proprietary systems. In Europe, ETSI (European Telecommunications Standards Institute) has been developing standards for satellite communication. The satellite standards are published under the DVB (Digital Video Broadcasting) project. ETSI is cooperating with the DVB consortium of hundreds of broadcasters, manufacturers, network operators, software developers, regulatory bodies, and others in many countries committed to designing global standards for the delivery of digital television and data services. Companies such as SES-ASTRA, Eutelsat, and Hispasat have been using the DVB standard.

Satellites can orbit the earth in several orbits referred to as GEO (Geosynchronous, or Geostationary Earth Orbit), MEO (Medium, or Middle Earth Orbit), and LEO (Low Earth Orbit). These satellites work in the Ku band (10–17 GHz frequency range) or the Ka band (18–31GHz frequency range).

GEO satellites orbit the earth at 22,300 miles above the earth's surface. They are tied to the earth's rotation and their positions are fixed in relation to earth's surface. This is an advantage since the earth station needs to point its transmitter and receiver to only one location in space to be able to transmit and receive from the GEO satellite. This makes GEO satellites popular for transmissions of high-speed data, television, and other wideband applications. A disadvantage is the long distance that the signal needs to travel. This results in a long time delay of few hundred millisecond, which causes deterioration in the QoS support provided for interactive applications that require short delays.

MEO satellites orbit the earth between 1,000 and 22,300 miles above the earth's surface. They are not stationary in relation to earth's surface. MEO satellites are used in geographical positioning systems.

LEO satellites orbit the earth between 400 and 1,000 miles above the earth's surface. LEOs are mostly used for data communication such as paging, email, and videoconferencing. LEO satellites do not have a fixed location in space in relation to the earth's surface. They move at very high speeds and to sustain end-to-end communication, they need to communicate with each other. Because they are located relatively close to earth, the earth station transmitter uses significantly less power. Also the travel time for the signal is significantly shorter, a fact that significantly decreases the multimedia application's delay, potentially improving QoS support.

13.1.1 DVB Return Channel System (DVB-RCS)

The DVB family of standards contains GEO satellite communication with various options of uplink (or return) and downlink (or forward) channels. Up to a few years ago, satellites were only able to transfer data unidirectionally from the content sources to the end user equipment, mostly TV sets. However, in recent years bidirectional communication has been introduced with various return channels such as telephones, ISDN, Cable TV, DECT, and LMDS. Here is a list of DVB standards:

- ETS 300 802, Digital Video Broadcasting (DVB); Network-independent protocols for DVB interactive services
- ETSI ES 200 800, Digital Video Broadcasting (DVB); DVB interaction channel for Cable TV distribution systems (CATV)
- TR 101 201, Technical Report Digital Video Broadcasting (DVB); Interaction channel for Satellite Master Antenna TV (SMATV) distribution systems; Guidelines for versions based on satellite and coaxial systems
- EN 301 193, European Standard (Telecommunications series) Digital Video Broadcasting (DVB); Interaction channel through the Digital Enhanced Cordless Telecommunications (DECT)
- EN 301 195, European Standard (Telecommunications series) Digital Video Broadcasting (DVB); Interaction channel through the Global System for Mobile communications (GSM)
- EN 301 199, European Standard (Telecommunications series) Digital Video Broadcasting (DVB); Interaction channel for Local Multi-point Distribution Systems (LMDS)

- ETS 300 801, Digital Video Broadcasting (DVB); Interaction channel through Public Switched Telecommunications Network (PSTN)/Integrated Services Digital Networks (ISDN)
- ETSI EN 301 958 V1.1.1 (2002-03) European Standard (Telecommunications series)
- Digital Video Broadcasting (DVB); Interaction channel for Digital Terrestrial Television (RCT) incorporating Multiple Access OFDM
- TR 101 194, Digital Video Broadcasting (DVB); Guidelines for implementation and usage of the specification of network-independent protocols for DVB interactive services

In this book we focus on the satellite DVB standards that provide two-way communication through the satellite, referred to as DVB-RCS. In these systems the end user has direct access to the satellite using the satellite's uplink and downlink channels. We focus on the following standards and documents:

- ETSI EN 301 790, European Standard (Telecommunications series), Digital Video Broadcasting (DVB); Interaction channel for satellite distribution systems
- ETSI TR 101 790, European Standard (Telecommunications series), Technical Report Digital Video Broadcasting (DVB); Interaction channel for Satellite Distribution Systems; Guidelines for the use of EN 301 790
- TR 101 202, European Standard (Telecommunications series), Digital Video Broadcasting (DVB); Implementation guidelines for Data Broadcasting
- EN 301 192, European Standard (Telecommunications series), Digital Video Broadcasting (DVB); DVB specification for data broadcasting
- ETS 300 802, European Standard (Telecommunications series), Digital Video Broadcasting (DVB); Network-independent protocols for DVB interactive services

The downlink related part of the specifications includes the DVB-Satellite (DVB-S) standard which is used in the typical satellite broadcast systems. For uplink communication, the DVB-RCS standard uses the RCST (Return Channel Satellite Terminal), a satellite terminal that supports interactive services such as SMATV (Satellite Master Antenna Television) as well as data transmission.

DVB-RCS MAC is based upon FDM/TDM, where the uplink and downlink transmissions take place at different frequency ranges. DVB-RCS does not define a specific operating frequency channel to be used, which allows flexible frequency deployment. Both the uplink and downlink channels are time slotted. The MAC specifications allow vendor specific implementations of various QoS supporting algorithms.

Depending on the dish size, DVB-RCS can achieve maximum data speeds as detailed in Table 13.1. The actual data speed depends on the DVB operator equipment and agreements with the end user.

Table 13.1   DVB-RCS Maximum Data Speeds

Dish Size	0.6 Meter	1.0 Meter	1.2 Meter
Return Link Speed	150 kbps	380 kbps	2 Mbps
Forward Link Speed	38 Mbps	38 Mbps	38 Mbps

13.2 Architecture

The DVB-RCS model is shown in Figure 13.1. DVB supports two channels: the Broadcast Channel and the Interaction Channel.

The Broadcast Channel, which is a unidirectional downlink broadcast channel, is identical to the channel defined in the satellite digital video broadcast (DVB-S) standard. The Interaction Channel provides bidirectional interaction communication between the service provider and the end-user. The Interaction Channel consists of a Forward Interaction channel (from the service provider to end-user) and Return Interaction channel (from the end-user to service provide). Typically, the Forward Interaction channel is included in the Broadcast Channel.

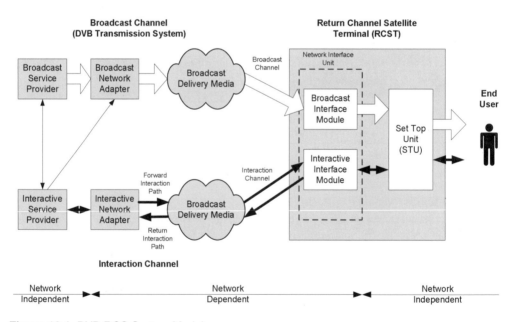

Figure 13.1 DVB-RCS System Model

The RCST provides interface for both Broadcast and Interaction Channels. It is supported by the Network Interface Unit and the Set Top Unit.

A satellite interactive network with several RCST channels is described in Figure 13.2. The NCC (Network Control Center) monitors and controls the operation of the satellite interactive network. The NCC manages the network resources and authorizes and allocates transmission resources to RCSTs. The system is fed by a few Feeder Stations. The Feeder Station on the forward link is a standard satellite digital video broadcast (DVB-S) on which we multiplex user data, control, and timing signals. The Traffic Gateway receives RCST return signals and provides accounting functions, interactive services, and connections to external public or private service providers. These service providers can be Internet data services, pay-per-view TV, financial services, and corporate networks.

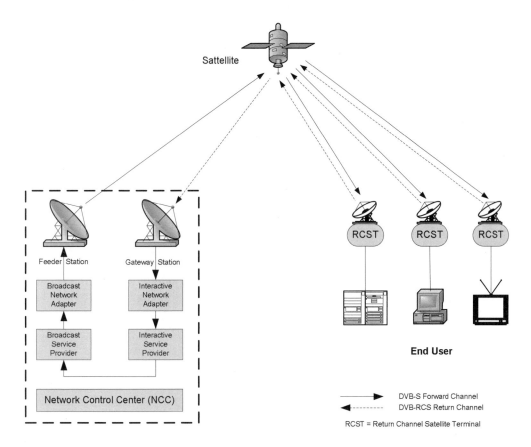

Figure 13.2 Overall Satellite Interactive Network Architecture

The forward link carries signaling information from the NCC and user traffic to RCSTs. The signaling traffic from the NCC to RCSTs, referred to as the Forward Link Signaling, includes the necessary information required to operate the return link. Both the user traffic and the forward link signaling can be carried over different forward link signals. Several RCST configurations are possible depending on the number of forward link receivers present on the RCST.

In the next sections we describe the forward and reverse link technologies that are used in the DVB-RCS standard.

13.3 Forward Link

The forward link or the broadcast channel is based on the DVB-S standard which employs MPEG-2 Transport Stream (TS) defined by ISO/IEC (International Standards Organization/ International Electrotechnical Commission) 13818-1 standard. Multiple data streams (i.e., video, audio, data) are multiplexed on MPEG-2 TS and delivered through the broadcast channel as shown in Figure 13.3. These data streams are categorized into several application areas as follows: DVB data piping, DVB data streaming, DVB multiprotocol encapsulation, DVB data carousel, and DVB object carousel. Each application area has different QoS requirements. Some data streams (i.e., DVB data piping, DVB data stream) are mapped directly onto MPEG-TS. Other data streams (i.e., DVB multiprotocol encapsulation, DVB data carousel, and DVB object carousel) are first mapped through Data Storage Media Command and Control (DSM-CC) and then MPEG-TS. The Forward Link employs Time Division Multiplex (TDM) on a single digital carrier. Before describing each application area, we would like to provide some background of MPEG-2 Transport Stream.

13.3.1 MPEG-2 Transport Stream

MPEG-2 Transport Stream is a transport format for video and audio transmission. Figure 13.4 illustrates a simplified overview of the MPEG-2 process.

The video encoder compresses the video signal and constructs video frames (i.e., I-frame, P-frame, B-frame). Each video frame includes the element stream (ES) header and a series of encoded video frames. The ES header provides information about the content and location of the video blocks required for decoding the video stream. The element streams are packetized and constructed into a packetized element stream (PES). Figure 13.5 shows the PES packet structure. The PES header contains the information necessary to decode the video stream. Important information in PES header that we would like to point out includes: 1) the Decode Time Stamp (DTS), which indicates when to decode the video frame and 2) the Presentation Time Stamp (PTS), which indicates when to display the video frame. PES packets from the different streams are multiplexed into a Transport Stream (TS). Figure 13.6 shows the Transport Stream packet structure. The packet ID (PID) in the TS header identifies the TS payload contents.

The list of PID and its associated values are included in the Program Specific Information (PSI) table. There are several types of PSI tables such as the Program Association Table (PAT), the Program Map Table (PMT), and the Conditional Access Table (CAT). Figure 13.7 illustrates the PSI table structure. Within the MPEG-2 Transport Stream, there are multiple programs (i.e., TV programs). Each program consists of multiple video streams, audio streams, and data streams. The PSI table provides the PID information required to find, identify, and reconstruct the video stream in subsequent Transport Stream packets. First, the receiver scans for the PAT identified by a reserved PID (PID = 0). PAT contains the mapping between the PID value and the associated program pointer of the PMT. For example, as shown in Figure 13.7, PID = 22 points to the PMT of program 1. The PMT contains the mapping between the PID value and the associated stream of the program. For example, as shown in Figure 13.7, PID = 62 identifies the video stream of program 1.

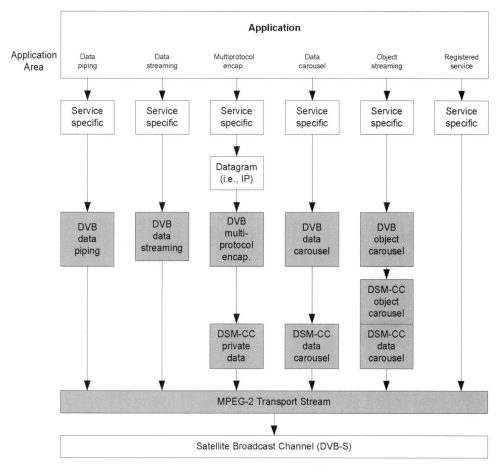

Figure 13.3 Data Broadcast Specification Overview (satellite side)

There are other types of tables such as the Superframe Composition Table (SCT), Frame Composition Table (FCT), and Time Slot Composition Table (TCT) used to define the channel structure. We will describe them later in this chapter.

13.3.2 Data Piping

The data piping mechanism uses an asynchronous (i.e., without timing, such as IP packets) transport mechanism. As shown in Figure 13.8, the data packets are inserted directly into the payload of the MPEG2 transport packets without any fragmentation and reassembly. At the TS header, the payload_unit_start_indicator field may be implemented to include the indication of the start of the data piping packet while the transport_priority field may be implemented to include the end of the packet.

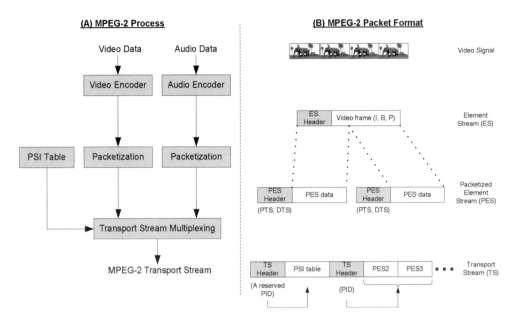

Figure 13.4 MPEG-2 Process Overview

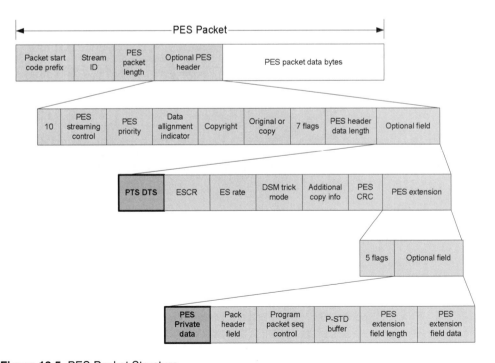

Figure 13.5 PES Packet Structure

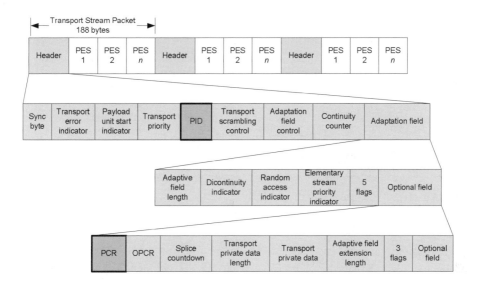

Figure 13.6 Transport Stream Packet Structure

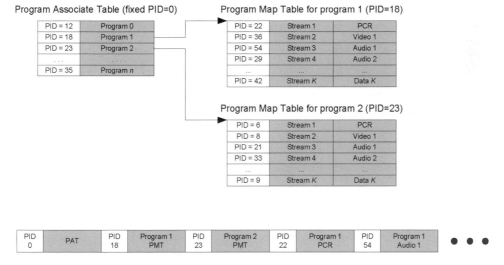

Figure 13.7 Program Specific Information (PSI) Table Structure

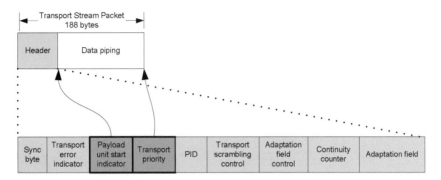

Figure 13.8 Data Piping Packets

13.3.3 Data Streaming

Data streaming mechanism is for streaming-oriented applications which may be asynchronous, synchronous, or synchronized data streams. Asynchronous data streaming transmits data without timing requirements while synchronous data streaming transmits data using a fixed rate transmission clock (i.e., T1, circuit emulation). Synchronized data streaming transmits data with timing related to the different kinds of data streams (i.e., streaming video along with audio where both require time synchronization).

As shown in Figure 13.9, the data streaming packets are inserted into the PES packet payload.

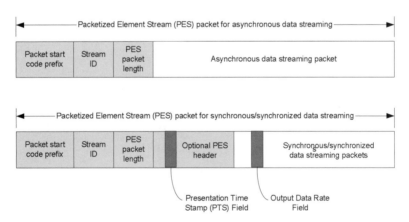

Figure 13.9 Packetized Element Stream (PES) Packets for Asynchronous/Synchronous/ Synchronized Data Streaming

The stream ID in the PES packet identifies the type of data streaming contained in the payload. The Presentation Time Stamp (PTS) described in Section 13.3.1 is an optional field in the PES header that includes time stamp information for synchronous and synchronized data streaming.

Packet scheduling algorithms can use the PTS information to determine when to transmit synchronous/synchronized packets. For synchronous data packets, there is a field in the PES payload called output data rate that includes the synchronous data stream required data rate.

13.3.4 Multiprotocol Encapsulation

Multiprotocol encapsulation method is used for transporting packets that originate from different network protocols on top of the MPEG2 Transport Streams in DVB networks. It has been optimized for delivery of IP packets but can be used for other protocols as well. A 48-bit MAC address is used to identify the receiver (i.e., RCST). UDP/IP or TCP/IP traffic will be multiplexed into an MPEG-2 transport stream through DSM-CC sections (defined in ISO/IEC 13818-6) as detailed in Figure 13.10. Using encryption, encapsulation can also provide secure transmission.

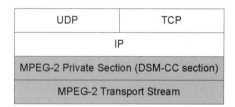

Figure 13.10 Mulitiprotocol Encapsulation for IP Traffic

The DSM-CC (Digital Storage Media Command and Control) facilitates the transmission of a structured group of objects from a broadcast server to clients using directory objects, file objects, and stream objects. It provides tools for controlling the MPEG-2 streams. DSM-CC may be used for controlling the video reception, providing features normally found on Video Cassette Recorders (VCRs) such as fast-forward, rewind, and pause. DSM-CC is designed for lightweight and fast operation considering limited memory devices.

The long delay imposed by the distances of thousands of miles that the signal needs to travel adversely affects the TCP features such as congestion control and packet acknowledgments. TCP's congestion control mechanism assumes that all data loss is due to congestion. Therefore, it wrongly translates the long delays to data loss and thus slows down the transmission of packets and prevents new connections. Moreover, the required TCP acknowledgments can congest the relative small bandwidth available on the uplink channel. Consequently, satellite vendors employ solutions that overcome such TCP problems. Such solutions may employ an intermediate server that acts as the end-user and projects timing parameters to the traffic source such that it will not slow down its transmissions. The intermediate server will encapsulate the packets in "new" packets and will transmit them to the end-user.

13.3.5 Data Carousel

The Data Carousel is a mechanism that allows a server to present a set of packets to a program, which is run by a receiver, by cyclically repeating the contents of the carousel one or more times. An example is the Teletext system in which a set of pages is cyclically broadcasted.

13.4 Return Link

The return link carries transmissions from the RCSTs to the satellite. As shown in Figure 13.11, the return channel is structured into superframes, frames, and time slots.

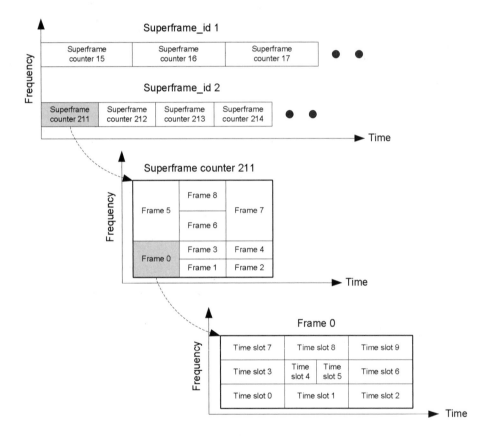

Figure 13.11 Superframes, Frames, and Timeslots

A superframe is defined by a section of frequency range and time duration of the return channel. Each superframe is labeled with a superframe_counter. A superframe consists of frames identified by frame_numbers. A frame consists of a time slot identified by timeslot_number. For channel allocation purposes, each time slot is uniquely identified by Superframe_id, Superframe_counter, Frame_number, and Timeslot_number. The superframe, frame, and time slot structure of the satellite system as well as their identifications are indicated in the Superframe Composition Table (SCT), the Frame Composition Table (FCT), and the Time Slot Composition Table (TCT), which are broadcasted by the satellite.

All RCST transmissions are governed and controlled by the NCC. Before the RCST can send data, it needs to join the network (logon) and inform the NCC about its configuration. The logon process occurs on a frequency channel that is defined by the NCC and transmitted on the Forward Link.

The Return Link supports four types of traffic: traffic (TRF), acquisition (ACQ), synchronization (SYNC), and common signaling channel (CSC).

TRF is used for carrying data from the RCST to the Gateway Station. This traffic can include ATM based cells or MPEG-2 packets. Synchronization and Acquisition traffic is used to position RCST transmissions during and after the logon process. A SYNC is used by an RCST to synchronize and send control information to the system. An ACQ can be used to achieve synchronization prior to operational use of the network by the RCST. The CSC packets are only used by the RCST to identify itself to the system during the logon process, as well as other set up information, such as determining the frequencies.

13.4.1 Return Link Channel Access

Because of the multipoint-to-point transmissions on the return link, the RCSTs employ a multiple access mechanism denoted Multifrequency Time Division Multiple Access (MF-TDMA). MF-TDMA employs a combination of FDMA and TDMA. In every superframe, the NCC broadcasts a Terminal Burst Time Plan (TBTP) which indicates the frequency, the bandwidth, the start time, and the duration for each RCST that is allowed to transmit data. There are two MF-TDMA schemes: fixed-slot MF-TDMA and dynamic-slot MF-TDMA. In fixed-slot MF-TDMA, the bandwidth and duration of the time slots assigned to an RCST are fixed while in dynamic-slot MF-TDMA, the bandwidth and duration of the time slots are variable. Dynamic-Slot MF-TDMA is optional in DVB systems. Figure 13.12 illustrates a channel access assignment diagram for a specific RCST.

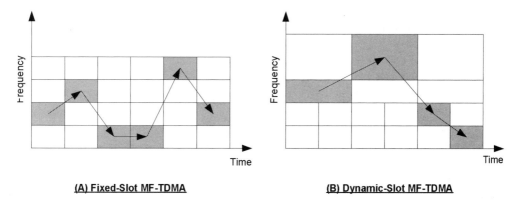

(A) Fixed-Slot MF-TDMA   (B) Dynamic-Slot MF-TDMA

Figure 13.12 Channel Access Assignment Diagram for a Specific RCST

The channel assignment is based on the application's bandwidth demand and QoS requirements. The channel assignment process can be achieved through a request-grant process where the RCST requests the bandwidth and the NCC grants the bandwidth. As shown in Figure 13.13, the standard defines capacity requests as well as signaling methods. The standard does not define the algorithm that determines the bandwidth allocation of each RCST.

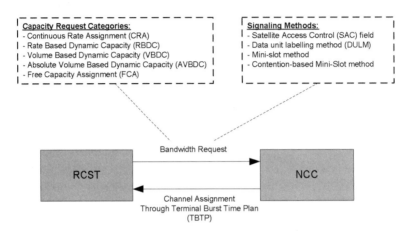

Figure 13.13 Capacity Request-Grant Process

Each application (identified by the Channel_ID) on a RCST requests bandwidth individually. The per-connection capacity request is categorized into the following five categories:

- Continuous Rate Assignment (CRA) indicates the fixed minimum bandwidth requirement per superframe. CRA is suitable for applications that require a fixed minimum guarantee rate with minimum delay and delay jitter such as constant bit rate (CBR) applications (circuit emulation).
- Rate-Based Dynamic Capacity (RBDC) indicates the dynamic bandwidth requirement per superframe. The current RBDC request will override all previous RBDC requests. To prevent a hanging capacity assignment, RBDC has a predetermined time-out period associated with it. If the time-out period expires, RBDC is set to zero. RBDC is suitable for variable bit rate (VBR) applications. Practically, the combination of CRA and RBDC can be used. CRA provides the fixed minimum bandwidth while RBDC provides dynamic bandwidth on the top of the minimum bandwidth.
- Volume-Based Dynamic Capacity (VBDC) indicates the volume bandwidth requirement per superframe. These requests are cumulative (i.e., each request increments on previous requests). At the end of each superframe, the cumulative amount of bandwidth is deducted by the amount of bandwidth already assigned in the superframe to the specific RCST. VBDC is suitable for delay tolerant applications.

- Absolute Volume-Based Dynamic Capacity (AVBDC) is similar to VBDC with the difference that AVBDC is not cumulative. It indicates the absolute value of the volume bandwidth. Therefore, the current AVBDC will override all previous AVBDCs. AVBDC is used in cases where VBDC might be lost during the request process.
- Free Capacity Assignment (FCA) uses the available bandwidth assignment. FCA is not mapped to any traffic category. The available bandwidth assignment is considered as bonus bandwidth used to reduce traffic delays.

The standard defines the following signaling methods that can be used for capacity requests:

- *Satellite Access Control (SAC) field:* A SYNC packet used for synchronization contains an optional SAC field. This field includes the connection capacity request. The SYNC packet can be used in contention mode where multiple RCSTs transmit SYNC packets at the same time.
- *Data Unit Labeling Method (DULM):* This method allows RCSTs to transmit control and management information to the NCC piggybacked to the data packets already sent during the assigned periods for TRF traffic.
- *Mini-slot Method:* This method is based on a periodic assignment to already logged-on RCSTs for packets that use less than full traffic time slots. It carries control and management information and is used also for maintaining RCST synchronization. This mechanism is supported by the SAC used in SYNC packets.
- *Contention-based Mini-slot Method:* This method can be applied such that a group of RCSTs can access a Mini-Slot on a contention basis. This mechanism is supported by the SAC request used in SYNC packets. The access is based on Slotted Aloha, in which the RCST transmits at the beginning of the time slot and if it is not successful it backs off and retransmits in another time slot.

13.5 Quality of Service Support

Figure 13.14 shows DVB-RCS QoS architecture. The standard defines some basic QoS mechanisms such as classification, channel access, and capacity request signaling.

13.5.1 Classification

Before packets from a specific application can be transmitted by the RCST, a connection needs to be established between the application and the NCC. Each application will be identified by the Channel_ID. DVB-RCS uses per-flow classification which enables per-flow QoS services. The signaling processes involved in the connection setup, modification, and termination are expected to be clarified in the next revision of the standard.

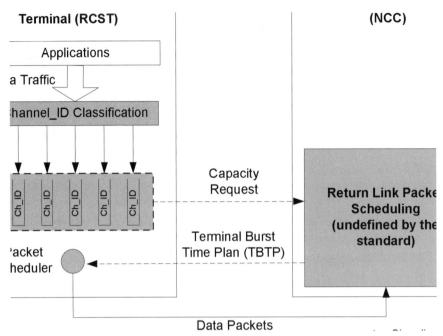

Figure 13.14 DVB-RCS QoS Architecture

13.5.2 Capacity Request Signaling and Channel Access

The standard defines the signaling method for capacity requests as well as the types of capacity requests. As described in Section 13.4.1 each type of capacity request which is per connection basis, supports different types of applications with different QoS requirements.

As described in Section 13.4.1, the return link employs MF-TDMA. Each RCST is allowed to transmit packets in a specific frequency and for a specific amount of time as defined in the Terminal Burst Time Plan (TBTP) as described in Section 13.4.1. TBTP is sent from the NCC to the RCST in every superframe. The TBTP mechanism enables dynamic bandwidth allocation which can provide QoS support for each application. The scheduling algorithm that determines the TBTP as a function of the application's QoS requirements as well as the available system resources is not included in the standard.

Acronyms and Abbreviations

Part 1: Multimedia Application and Quality of Service (QoS)

CBR	Constant Bit Rate
CSMA/CA	Carrier Sense Multiple Access with Collision Avoidance
CSMA/CD	Carrier Sense Multiple Access with Collision Detection
DiffServ	Differentiated Service
DSCP	Differential Service Code Point
EMS	Enhanced Messaging Service
GPS	Global Positioning System
IntServ	Integrated Service
MMS	Multimedia Messaging Service
MOS	Mean Opinion Score
QoS	Quality of Service
RSVP	Resource Reservation Protocol
SBM	Subnet Bandwidth Manager
SMS	Short Message Service

TDMA   Time Division Multiple Access
TOS   Type Of Service

VBR   Variable Bit Rate
VOD   Video On Demand
VoIP   Voice over IP

WAP   Wireless Application Protocol
WFQ   Weight Fair Queue

Part 2: Wireless Local Area Networks

ACF   Association Control Function
ACH   Access Feedback Channel
AIFS   Arbitration Interframe Space
AP   Access Point
ARQ   Automatic Repeat Request
ASCH   Association Control Channel

BC   Broadcast
BCCH   Broadcast Control Channel
BCH   Broadcast Channel
BSS   Basic Service Set

CC   Controlled Contention
CCA   Clear Channel Assessment
CCI   Controlled Contention Interval
CFP   Contention-Free Period
CM   Centralized Mode
CP   Contention Period, Connection Point
CTS   Clear To Send
CW   Contention Window

DCC   DLC Connection Control
DCCH   Dedicated Control Channel
DCF   Distributed Coordination Function
DFS   Dynamic Frequency Selection
DIFS   DCF Interframe Space
DiL   Direct Link
DL   DownLink
DLC   Data Link Control

DM	Direct Mode
DS	Distribution System
DSSS	Direct Sequence Spread Spectrum
EC	Error Control
EDCF	Enhanced DCF
EIFS	Extended Interframe Space
ESS	Extended Service Set
FCCH	Frame Control Channel
FCH	Frame Channel
FHSS	Frequency Hopping Spread Spectrum
HC	Hybrid Coordinator
HCF	Hybrid Coordination Function
IBSS	Independent Basic Service Set
IFS	Interframe Space
LCCH	Link Control Channel
LCH	Long Transport Channel
MT	Mobile Terminal
NAV	Network Allocation Vector
OFDM	Orthogonal Frequency Division Multiplexing
PCF	Point Coordination Function
PF	Persistence Factor
PIFS	PCF Interframe Space
PLCP	Physical Layer Convergence Protocol
QSTA	QoS Station
RA	Random Access
RBCH	RLC Broadcast Channel
RCH	Random Access Channel
RFCH	Random Access Feedback Channel
RG	Resource Grants

RLC	Radio Link Control
RR	Resource Request
RRC	Radio Resource Control
RTS	Request To Send
SCH	Short Transport Channel
SIFS	Short Interframe Space
TC	Traffic Category
TS	Traffic Stream
TSPEC	Traffic Specifications
TXOP	Transmission Opportunity
UBCH	User Broadcast Channel
UDCH	User Data Channel
UL	Uplink
UMCH	User Multicast Channel
UWB	Ultra Wide Band

Part 3: Wireless Metropolitan Area Networks

BE	Best Effort
BS	Base Station
CID	Connection Identifier
CPS	Common Part Sublayer
CS	Convergence Sublayer
DCD	Downlink Channel Descriptor
DIUC	Downlink Interval Usage Code
GPC	Grant Per Connection
GPSS	Grant Per Subscriber Station
IE	Information Element
IUC	Interval Usage Code
nrtPS	Non-Real-Time Polling Service
PMP	Point to Multipoint

rtPS   Real-Time Polling Service

SI   Slip Indicator
SS   Subscriber Station

UCD   Uplink Channel Descriptor
UGS   Unsolicited Grant Service
UIUC   Uplink Interval Usage Code

Part 4: Wireless Personal Area Networks

ACL   Asynchronous Connection-Less

CAC   Channel Access Code
CAP   Contention Access Period
CFP   Contention-Free Period
CTA   Channel Time Allocation
CTR   Channel Time Request

DAC   Device Access Code

FHS   Frequency Hop Synchronization

GTS   Guaranteed Time Slot

HCI   Host Control Interface
HID   Human Interface Device

IAC   Inquiry Access Code

L2CAP   Logical Link Control and Adaptation Protocol
LC   Link Control
LM   Link Manager

PNC   PicoNet Coordinator
PNID   PicoNet ID

MTS   Management Time Slot

SCO   Synchronous Connection-Oriented
SDP   Service Discovery Protocol

UA	User Asynchronous
UI	User Isochronous
US	User Synchronous

Part 5: 2.5G and 3G Networks

GPRS, UMTS, cdma2000

3GPP	3G Partnership Project
3GPP2	3G Partnership Project 2
AAA	Authorization And Accounting
AMC	Adaptive Modulation and Coding
AMR	Adaptive Multirate
ARIB	Association of Radio Industries and Businesses
ARQ	Automatic Repeat Request
ASC	Access Service Classes
BCCH	Broadcast Control Channel
BCH	Broadcast Channel
BMC	Broadcast/Multicast Control
BSC	Base Station Controller
BSS	Base Station Subsystem
BTS	Base Transceiver Station
CCCH	Common Control Channel
CCH	Control Channel
CCITT	Consultative Committee for International Telephony and Telegraphy
CDMA	Code Division Multiple Access
CPAGCH	Compact Packet Access Grant Channel
CPBCCH	Compact Packet Broadcast Control Channel
CPCH	Common Packet Channel
CPNCH	Compact Packet Notification Channel
CPPCH	Compact Packet Paging Channel
CPRACH	Compact Packet Random Access Channel
CSCF	Call State Control Function
CTCH	Common Traffic Channel
CWTS	China Wireless Telecommunication Standard Group
DBPSCH	Dedicated Basic Physical Subchannel
DCCH	Dedicated Control Channel

DCH	Dedicated Channel
DL	Data Link
DSCH	Downlink Shared Channel
DTCH	Dedicated Traffic Channel
DTM	Dual Transfer Mode
EDGE	Enhanced Data Rate for Global Evolution
EGPRS	Enhanced GPRS
ETSI	European Telecommunications Standards Institute
FACH	Forward Access Channel
F-APICH	Forward Dedicated Auxiliary Pilot Channel
F-ATDPICH	Forward Auxiliary Transmit Diversity Pilot Channel
F-BCCH	Forward Broadcast Control Channel
F-CACH	Forward Common Assignment Channel
F-CPCCH	Forward Common Power Control Channel
F-PCH	Forward Paging Channel
F-PDCCH	Forward Packet Data Control Channel
F-PDCH	Forward Packet Date Channel
F-QPCH	Forward Quick Paging Channel
FR	Full Rate
F/R-CCCH	Forward/Reverse Common Control Channel
F/R-DCCH	Forward/Reverse Dedicated Control Channel
F/R-FCH	Forward/Reverse Fundamental Channel
F/R-PICH	Forward/Reverse Pilot Channel
F/R-SCCH	Forward/Reverse Supplemental Code Channel
F/R-SCH	Forward/Reverse Supplemental Channel
F-SYNCH	Forward Synchronous Channel
F-TDPICH	Forward Transmit Diversity Pilot Channel
GERAN	GSM/EDGE Radio Access Network
GGSN	Gateway GPRS Support Node
GPRS	General Packet Radio Service
G-RNTI	GERAN Radio Network Temporary Identity
GSM	Global System for Mobile Communications
HARQ	Hybrid Automatic Repeat Request
HLR	Home Location Register
HR	Half Rate
HS	High-Speed

HSCSD	High-Speed Circuit Switched Data
HS-DSCH	High-Speed Downlink Shared Channel
HSDPA	High-Speed Downlink Packet Access
HSPDA	High-Speed Download Packet Access
HSS	Home Subscriber Server
IMS	Internet Multimedia Subsystem
IMT-2000	International Mobile Telecommunication–2000
ITU	International Telecommunications Union
LAC	Link Access Control
MAC	Medium Access Control
MGCF	Media Gateway Control Function
MGW	Media Gateway
MRF	Multimedia Resource Function
MS	Mobile Station
MSC	Mobile Switching Center
OVSF	Orthogonal Variable Spreading Factor
PACCH	Packet Associated Control Channel
PAGCH	Packet Access Grant Channel
PBCCH	Packet Broadcast Control Channel
PCCCH	Packet Common Control Channel
PCH	Paging Channel
PCM	Pulse Code Modulation
PDCCH	Packet Dedicated Control Channel
PDCH	Packet Data Channel
PDCHCF	Packet Data Channel Control Function
PDCP	Packet Data Convergence Protocol
PDP	Packet Data Protocol
PDSN	Packet Data Serving Node
PDTCH	Packet Data Traffic Channel
PNCH	Packet Notification Channel
PPCH	Packet Paging Channel
PRACH	Packet Random Access Channel
PSTN	Public Switched Telephone Network
PTCCH/D	Packet Timing Advance Control Channel/Downlink
PTCCH/U	Packet Timing Advance Control Channel/Uplink

PTM	Point-To-Multipoint
PTM-M	Point-To-Multipoint–Multicast
PTP	Point-To-Point
QoS	Quality of Service
RACH	Random Access Channel
R-ACH	Reverse Access Channel
R-ACKCH	Reverse Acknowledgment Channel
RAN	Radio Access Network
RCN	Radio Network Controller
R-CQICH	Reverse Channel Quality Indicator Channel
R-EACH	Reverse Enhanced Access Channel
RLC	Radio Link Control
RLP	Radio Link Protocol
RNC	Radio Network Controller
RR	Radio Resource
RRC	Radio Resource Control
R-SGW	Roaming Signaling Gateway
RX	Receive
SBPSCH	Shared Basic Physical Subchannel
SF	Spread Factor
SGSN	Serving GPRS Support Node
SHCCH	Shared Channel Control Channel
SRBP	Signaling Radio Burst Protocol
SS7	Signaling System No. 7
TBF	Temporary Block Flow
TCH	Traffic Channel
TDMA	Time Division Multiple Access
TE	Terminal Equipment
TFC	Transport Format Combination
TFCS	Transport Format Combination Set
TFI	Transport Format Indicator, Temporary Flow Identity
TIA/EIA	Telecommunications Industry Alliance/Electronics Industries
TLLI	Temporary Logical Link Identifier
TTA	Telecommunication Technology Association
TTI	Transmission Time Interval
T-SGW	Transport Signaling Gateway

TX   Transmit

UE   User Equipment
UMTS   Universal Mobile Telecommunications System
USCH   Uplink Shared Channel
USF   Uplink State Flag
UTRAN   UMTS Terrestrial Radio Access Network

VLR   Visitor Location Register

WCDMA   Wide CDMA or Wide Code Division Multiple Access

Satellite Communication

ACQ   Acquisition
AVBDC   Absolute Volume Based Dynamic Capacity

CRA   Continuous Rate Assignment
CSC   Common Signaling Channel

DSM-CC   Digital Storage Media Command and Control
DTS   Decode Time Stamp
DULM   Data Unit Labeling Method
DVB   Digital Video Broadcasting Project
DVB-S   Digital Video Broadcast by Satellite

FCA   Free Capacity Assignment

GEO   Geosynchronous or Geostationary Earth Orbit

ISO   International Standards Organization

LEO   Low Earth Orbit

MEO   Medium or Middle Earth Orbit

NCC   Network Control Center

PAT   Program Association Table
PES   Packetized Element Stream
PID   Packet ID

PMT	Program Map Table
PSI	Program-Specific Information
PTS	Presentation Time Stamp
QPSK	Quadrature Phase Shift Keying
RBDC	Rate-Based Dynamic Capacity
RCST	Return Channel Satellite Terminal
SAC	Satellite Access Control
SMATV	Satellite Master Antenna Television
SYNC	Synchronization
TBTP	Terminal Burst Time Plan
TRF	Traffic
TS	Traffic Stream
VBDC	Volume-Based Dynamic Capacity

Bibliography

Introduction

R. D. Nee and R. Prasad, *OFDM for Wireless Multimedia Communications*, Artech House, Norwood, MA, 2000.

M. K. Simon, J. K. Omura, R. A. Scholtz, and B. K. Levitt, *Spread Spectrum Communications Handbook*, McGraw-Hill Professional, New York, 2001.

R. Stevens, *The Protocols (TCP/IP Illustrated, Volume 1)*, Addison-Wesley, Boston, 1994.

A. Tanenbaum, *Computer Networks*, Prentice Hall, Upper Saddle River, NJ, 2002.

Part 1: Multimedia Applications and Quality of Service (QoS)

RFC (Request for Comments, Internet RFC/STD/FYI/BCP Archives) 1349, P. Almquist, "Type of Service in the Internet Protocol Suite," July 1992.

RFC 1633, R. Braden, D. Clark, and S. Shenker, "Integrated Services in the Internet Architecture: an Overview," June 1994.

RFC 2205, R. Braden, L. Zhang, S. Berson, S. Herzog, and S. Jamin, "Resource ReSerVation Protocol (RSVP)," September 1997.

RFC 2211, J. Wroclawski, "Specification of the Controlled-Load Network Element Service," September 1997.

RFC 2212, S. Shenker, C. Partridge, and R. Guerin, "Specification of Guaranteed Quality of Service," September 1997.

RFC 2474, K. Nichols, S. Blake, F. Baker, and D. Black, "Definition of the Differentiated Services Field (DS Field) in the IPv4 and IPv6 Headers," December 1998.

RFC 2475, S. Blake, D. Black, M. Carlson, E. Davies, Z. Wang, and W. Weiss, "An Architecture for Differentiated Services," December 1998.

RFC 2814, R. Yavatkar, D. Hoffman, Y. Bernet, F. Baker, and M. Speer, "SBM (Subnet Bandwidth Manager): A Protocol for RSVP-Based Admission Control over IEEE 802-Style Networks," May 2000.

Y. Bernet, *Networking Quality of Service and Windows Operating Systems,* Que Publishing, Indianapolis, IN, 2001.

A. Croll and E. Packman, *Managing Bandwidth: Deploying QoS in Enterprise Networks,* Prentice Hall, Upper Saddle River, NJ, 2000.

A. Demers, S. Keshav, and S. Shenker, *"Analysis and Simulation of a Fair Queuing Algorithm,"* SIG-COMM CCR 19, no. 4 (1989).

S. Floyd and V. Jacobson, *"Link-Sharing and Resource Management Models for Packet Networks,"* *IEEE/ACM Transactions on Networking 3,* no. 4, 1995.

F. Fluckiger, *Understanding Networked Multimedia,* Prentice Hall, Upper Saddle River, NJ, 1995.

D. Miras, *A Survey on Network QoS Needs of Advance Internet Applications (Working Document),* Internet2 QoS Working Group, December 2002. Available at http://qos.internet2.edu/wg/apps/fellow-ship/Docs/Internet2AppsQoSNeeds.html.

S. V. Raghavan and S. K., Tripathi, *Networked Multimedia Systems: Concepts, Architecture, and Design,* Prentice Hall, Upper Saddle River, NJ, 1998.

R. Steinmetz, and K. Nahrstedt, *Multimedia: Computing, Communications and Applications,* Prentice Hall, Upper Saddle River, NJ, 1995.

Part 2: Wireless Local Area Networks

IEEE 802.11 WG, ANSI/IEEE Std 802.11, "Information technology—Telecommunications and information exchange between systems—Local and metropolitan area networks—Specific Requirements—Part 11: Wireless LAN Medium Access Control (MAC) and Physical Layer (PHY) Specifications," 1999.

IEEE 802.11 WG, IEEE Std 802.11a-1999, "Supplement to IEEE Standard for Information Technology—Telecommunications and information exchange between systems—Local and metropolitan area networks—Specific requirements—Part 11: Wireless LAN Medium Access Control (MAC) and Physical Layer (PHY) specifications: High-speed Physical Layer in the 5 GHz Band," 1999.

IEEE 802.11 WG, IEEE Std 802.11b-1999, "Supplement to IEEE Standard for Information Technology—Telecommunications and information exchange between systems—Local and metropolitan area networks—Specific requirements—Part 11: Wireless LAN Medium Access Control (MAC) and Physical Layer (PHY) specifications: Higher-speed Physical Layer Extension in the 2.4 GHz Band," 1999.

IEEE 802.11 WG, IEEE Std 802.11e/D3.0,"Draft Supplement to STANDARD for Telecommunications and Information Exchange Between Systems—LAN/MAN Specific Requirements—Part 11: Wireless Medium Access Control (MAC) and Physical Layer (PHY) specifications: Medium Access Control (MAC) Enhancements for Quality of Service (QoS)," May 2002.

IEEE 802.11 WG, IEEE Std 802.11g/D2.1, "Draft Supplement to IEEE Standard for Information Technology—Telecommunications and information exchange between systems—Local and metropolitan area networks—Specific requirements—Part 11: Wireless LAN Medium Access Control (MAC) and Physical Layer (PHY) specifications: Further Higher-speed Physical Layer Extension in the 2.4 GHz Band," January 2002.

ETSI BRAN, TR 101 301, "Broadband Radio Access Network (BRAN); High Performance Radio Local Area Network (HIPERLAN) Type 2; Requirements and architecture for wireless broadband access," January 1999.

ETSI BRAN, ETSI TR 101 683, "Broadband Radio Access Network (BRAN); HIPERLAN Type 2; System Overview," February 2000.

ETSI BRAN, ETSI TR 101 761-1, "Broadband Radio Access Network (BRAN); HIPERLAN Type 2; Data Link Control (DLC) Layer; Part1: Basic Data Transport Functions," December 2001.

ETSI BRAN, ETSI TR 101 761-2, "Broadband Radio Access Network (BRAN); HIPERLAN Type 2; Data Link Control (DLC) Layer; Part2: Radio Link Control (RLC) sublayer," January 2002.

HomeRF Technical Committee, "HomeRF Specification Revision 2.0 Draft 20010507," May 2001.

Part 3: Wireless Metropolitan Area Networks

IEEE 802.16 WG, IEEE 802.16-2001, "IEEE Standard for Local and Metropolitan Area Networks—Part 16: Air Interface for Fixed Broadband Wireless Access Systems," April 2002.

IEEE 802.16 WG, IEEE P802.16a/D4-2002, "Draft Amendment to IEEE Standard for Local and Metropolitan Area Networks—Part 16: Air Interface for Fixed Broadband Wireless Access Systems—Medium Access Control Modifications and Additional Physical Layer Specifications for 2-11 GHz," May 2002.

Eklund, C., R. B. Marks, K. L. Standwood, and S. Wang, "IEEE Standard 802.16: A Technical Overview of the WirelessMAN Air Interface for Broadband Wireless Access," *IEEE Communication Magazine,* June 2002, pp. 98-107.

Part 4: Wireless Personal Area Networks

Bluetooth Specification version 1.1 volume 1, "Specification of the Bluetooth System: Core," February 2001.

Bluetooth Specification version 1.1 volume 2, "Specification of the Bluetooth System: Profiles," February 2001.

IEEE 802.15 WG, IEEE P802.15.1/D1.0.1, "Draft Standard for Information Technology—Telecommunications and information exchange between systems—Local and metropolitan area networks—Specific requirements—Part 15.1: Wireless Medium Access Control (MAC) and Physical Layer (PHY) specifications for Wireless Personal Area Networks (WPANs)," September 2001.

IEEE 802.15 WG, IEEE P802.15.3/D10, "Draft Standard for Telecommunications and Information Exchange Between Systems—LAN/MAN Specific Requirements—Part 15: Wireless Medium Access Control (MAC) and Physical Layer (PHY) specifications for High Rate Wireless Personal Area Networks (WPAN)," June 2002.

IEEE 802.15 WG, IEEE P802.15.4/D13, "Draft Standard for Telecommunications and Information Exchange Between Systems—LAN/MAN Specific Requirements—Part 15.4: Wireless Medium Access Control (MAC) and Physical Layer (PHY) specifications for Low Rate Wireless Personal Area Networks (LR-WPANs)," December 2001.

Part 5: 2.5G and 3G Networks

GPRS

3GPP TS 22.060, "3rd Generation Partnership Project; Technical Specification Group Services and System Aspects; General Packet Radio Service (GPRS); Service description, Stage 1 (Release 5)," March 2002.

3GPP TR 22.941, "3rd Generation Partnership Project; Technical Specification Group Services and System Aspects; IP Based Multimedia Services Framework, Stage 0 (Release 5) TS 22.228 3rd Generation Partnership Project; Technical Specification Group Services and System Aspects; Service requirements for the IP Multimedia Core Network Subsystem, Stage 1 (Release 5)," November 2001.

3GPP TS 43.051, "3rd Generation Partnership Project; Technical Specification Group GSM/EDGE Radio Access Network; Overall description, Stage 2 (Release 5)," April 2002.

3GPP TS 43.064, "3rd Generation Partnership Project; Technical Specification Group GERAN; Digital cellular telecommunications system (Phase 2+); General Packet Radio Service (GPRS); Overall description of the GPRS radio interface, Stage 2 (Release 5)," April 2002.

3GPP TS 44.118, "3rd Generation Partnership Project; Technical Specification Group GSM EDGE Radio Access Network; Mobile radio interface layer 3 specification, Radio Resource Control (RRC) Protocol, Iu Mode (Release 5)," April 2002.

3GPP TS 44.018, "3rd Generation Partnership Project; Technical Specification Group GSM/EDGE Radio Access Network; Mobile radio interface layer 3 specification; Radio Resource Control Protocol (Release 5)," April 2002.

3GPP TS 44.060, "3rd Generation Partnership Project; Technical Specification Group GSM/EDGE Radio Access Network; General Packet Radio Service (GPRS); Mobile Station (MS) - Base Station System (BSS) interface; Radio Link Control/Medium Access Control (RLC/MAC) protocol (Release 5)," May 2002.

3GPP TS 44.160, "3rd Generation Partnership Project; Technical Specification Group GSM/EDGE Radio Access Network; Mobile Station (MS)—Base Station System (BSS) interface; Radio Link Control/ Medium Access Control (RLC/MAC) protocol; Iu mode (Release 5)," March 2002.

3GPP TS 45.005, "3rd Generation Partnership Project; Technical Specification Group GSM/EDGE Radio Access Network; Radio transmission and reception (Release 5)," April 2002.

UMTS

3GPP TS 23.002, "3rd Generation Partnership Project; Technical Specification Group Services and Systems Aspects; Network architecture (Release 5)," June 2002.

3GPP TS 25.211, "3rd Generation Partnership Project; Technical Specification Group Radio Access Network; Physical channels and mapping of transport channels onto physical channels (FDD) (Release 5)," June 2002.

3GPP TS 25.301, "3rd Generation Partnership Project; Technical Specification Group Radio Access Network; Radio Interface Protocol Architecture (Release 5)," June 2002.

3GPP TS 25.302, "3rd Generation Partnership Project; Technical Specification Group Radio Access Network; Services provided by the physical layer (Release 5)," December 2002.

3GPP TS 25.308, "3rd Generation Partnership Project; Technical Specification Group Radio Access Network; High Speed Downlink Packet Access (HSDPA); Overall description, Stage 2 (Release 5)," March 2002.

3GPP TS 25.321, "3rd Generation Partnership Project; Technical Specification Group Radio Access Network; MAC protocol specification (Release 5)," June 2002.

3GPP TS 25.322, "3rd Generation Partnership Project; Technical Specification Group Radio Access Network; Radio Link Control (RLC) protocol specification (Release 5)," June 2002.

3GPP TS 25.323, "3rd Generation Partnership Project; Technical Specification Group Radio Access Network; Packet Data Convergence Protocol (PDCP) Specification (Release 5)," June 2002.

3GPP TS 25.331, "3rd Generation Partnership Project; Technical Specification Group Radio Access Network; Radio Resource Control (RRC); Protocol Specification (Release 5)," June 2002.

3GPP TS 25.401, "3rd Generation Partnership Project; Technical Specification Group Radio Access Network; UTRAN Overall Description (Release 5)," June 2002.

cdma2000

3GPP2 C.S0001-C, "Introduction to cdma2000 Standards for Spread Spectrum Systems Release C," May 2002.

3GPP2 C.S0002-C, "Physical Layer Standard for cdma2000 Spread Spectrum Systems Release C," May 2002.

3GPP2 C.S0003-C, "Medium Access Control (MAC) Standard for cdma2000 Spread Spectrum Systems Release C," May 2002.

3GPP2 C.S0004-C, "Signaling Link Access Control (LAC) Standard for cdma2000 Spread Spectrum Systems Release C," May 2002.

3GPP2 C.S0005-C, "Upper Layer (Layer 3) Signaling Standard for cdma2000 Spread Spectrum Systems Release C," May 2002.

Satellite Communication

ETSI EN 301 790, "European Standard (Telecommunications series), Digital Video Broadcasting (DVB); Interaction channel for satellite distribution systems," December 2000.

ETSI TR 101 790, "European Standard (Telecommunications series), Technical Report Digital Video Broadcasting (DVB); Interaction channel for Satellite Distribution Systems; Guidelines for the use of EN 301 790," September 2001.

ETSI TR 101 202, "European Standard (Telecommunications series), Digital Video Broadcasting (DVB); Implementation guidelines for Data Broadcasting," February 1999.

ETSI EN 301 192, "European Standard (Telecommunications series), Digital Video Broadcasting (DVB); DVB specification for data broadcasting," January 2003.

ETSI ETS 300 802, "European Standard (Telecommunications series), Digital Video Broadcasting (DVB); Network-independent protocols for DVB interactive services," November 1997.

ISO/IEC 13818-1, "Information Technology—Generic Coding of Moving Pictures and Associated Audio: Systems, Recommendation H.220.0," November 1994.

INDEX